newmedia.com.au

The changing face of Australia's media and communications

Trevor Barr

ALLEN & UNWIN

For Jacob and Luke

Copyright © Trevor Barr 2000

All rights reserved. No part of this book may be reproduced or transmitted in any form or by any means, electronic or mechanical, including photocopying, recording or by any information storage and retrieval system, without prior permission in writing from the publisher.

First published in 2000
Allen & Unwin
9 Atchison Street, St Leonards NSW 1590 Australia
Phone: (61 2) 8425 0100
Fax: (61 2) 9906 2218
E-mail: frontdesk@allen-unwin.com.au
Web: http://www.allen-unwin.com.au

National Library of Australia
Cataloguing-in-Publication entry:

Barr, Trevor, 1940– .
 newmedia.com.au: the changing face of Australia's
 media and communications.

 Bibliography.
 Includes index.
 ISBN 1 86508 049 7.

 1. Mass media—Australia. 2. Mass media policy—
 Australia. 3. Information technology—Australia.
 4. Telecommunication—Australia. 5. Telecommunication
 policy—Australia. 6. Telecommunication policy.
 7. Competition, International. I. Title.

302.230994

Set in 11/13 pt Bembo by DOCUPRO, Sydney
Printed by Griffin Press Pty Ltd, South Australia

Contents

Acknowledgments	vii
Preface	viii

1 Media moguls: Power, ownership and influence   **1**
Elites: Concentration of media ownership and control   5
Broadcasting licences: Partisan patronage   10
The old media hegemony   16

2 Forces for change: Communications as catalyst   **20**
Technological processes: Convergence and digitalisation   22
Communications in the global marketplace   31
Globalisation, the nation state and communications   34

3 Challenge of change: Australia's media institutions   **40**
Major media: The commercial club   41
Adding a wheel: Pay television in Australia   58
Identity revisited: Public broadcasting   61

4 Citizens to customers: Telecommunications in transition   **75**
A network for the nation   78
Conflicting interests, conflicting objectives   86
Deregulation: The awesome foursome   90
Telstra: Sale of the century   105
Australian telecommunications: Reviewing the new balance sheet   113

5 Electronic nomads: Internet as paradigm   **117**
Origins and growth of the Internet   120
Cyberspace and virtual communities   123
Changing place and space   128
Governance of the Internet: Censorship and privacy   130
New capitalism of convergence: The Internet economy   135

6 Being human: Paradoxes of the new media	**145**
Paradox 1: The paradox of equity	147
Paradox 2: The paradox of plenty	151
Paradox 3: The paradox of users	156
Paradox 4: The paradox of diversity	160
7 Towards an information society	**166**
International communications development policies	169
Australia's convergent communications economy	176
Australia as an IT client state	180
8 Whose vision?: Third way communications	**189**
The third communications order: e-communities/ e-Commerce	191
Universal service opportunities	197
Digital *déjà vu*	203
9 Re-thinking our communications strategy	**209**
Politics, not policies	210
Discourses for development	213
Towards an Australian communications policy blueprint	215
Whither Australia's communications?	234
Endnotes	*236*
Select bibliography	*249*
Index	*257*

Acknowledgments

My fourth book in this field has been the most demanding of all to research and conceptualise, but by far the most rewarding. The big picture is never easy to try to construct, let alone trying to take others along on the journey.

This work emerged in association with the development of a Master of Communications course in the School of Social and Behavioural Sciences at Swinburne University of Technology in Melbourne, and was later part of a Swinburne initiative to create an Institute of Social Research (ISR). Many people have made a valuable contribution to this work, and for undertaking the painstaking task of reading and offering criticism of the draft manuscript I am indebted to Mike Bowen, Maureen le Blanc, Peter Gerrand, Liz Harper, Carole Hooper, Joanne Jacobs, Ross Kelso, Ann Knowles, Dianne Northfield, Kevin Patrick, John de Ridder and John Schwartz. The research support of Carole Hooper and Youngmi Choi is also gratefully acknowledged.

I am most indebted to Allen & Unwin staff, especially Elizabeth Weiss and Colette Vella for their publishing advice and support, and to the editorial work of Michael Wall and Robyn Flemming.

In a world where commentary has been so captured by vested interests, the author ought to declare that he has *no* shareholding in any of the companies analysed in this text. In fact, he holds no shares at all. If, however, the notion of citizenry has any meaning in the future, then all Australians are shareholders in our national communications future.

Preface

In the burgeoning field of communications there is a great need for challenging, accessible and multiple sources of interpretation. The conceptual framework offered in this book is intended to enable readers to understand the exponential growth and convergence of media, information technology and telecommunications, collectively referred to as *communications*. No longer should communications be merely examined as an industry, but rather as a phenomenon which has now infiltrated virtually every aspect of our lives and which affects how society is organised.

Although Australia is the focal point of this analysis, the issues in communications are set within the new global framework. Clearly, we are living through an age of unprecedented developments in communications technologies and in the provision of new information services. Communications has become everybody's favourite growth business, globally. Among academics, economists, consultants, merchant bankers and business information personnel there is much disagreement about the scope and boundaries of what constitutes an information-based economy, and problems associated with attempts to construct and analyse data about its size and growth, both domestically and internationally. Yet, despite the difficulties experienced in attempting to obtain a 'full picture' of the global information economy, a substantial amount of data is available about the growth of particular goods and services. Some key indicators are:

- By 1996 about 40 per cent of households worldwide had a telephone.[1]
- Between 1992 and 1997 there was a twenty-fold increase globally in the number of Internet hosts.[2]
- In 1994 Internet traffic in Australia was only 1 per cent of the size of voice traffic on international links. By 1998 it had grown by

10 per cent per month—100 times in four years—to exceed voice traffic on these links.[3]
- The volume of international telephone traffic rose on average by a remarkable 15 per cent per annum between 1975 and 1995.[4]

What seems possible today in communications was unimaginable only two decades ago. Technological developments such as high-capacity information networks, the Internet and the World Wide Web, pay and digital television, and on-line services, are not merely electronic innovations which provide new technical tools or techniques and great new business opportunities. They strike at the heart of human decision-making about the acceptance and place of new technologies in our lives. Contemporary society is now commonly labelled as *the information society*—information has become the commodity of the future. Yet, if we are living through a communications revolution, who are the revolutionaries?

Our media columns and technology programs continually offer us a new array of communications gadgetry and slicker services to do things more quickly—but these are not the focus of this study. The primary framework of this book is to situate the major changes in communications within the field of political economy, and institutional and policy studies. The book sites the alleged communications revolution within the contemporary international economy, notably the role of communications in globalisation, and the drive for a more privatised, deregulated, information-based Australian economy. The fundamental questions explored are: *who and what is driving these changes, and how; whose interests are being served; and who benefits?* The search for answers to such questions is conceptually inseparable from the political economy and from attempting to understand how society works.

For the purposes of this book, the term 'communications' is used in an encompassing sense. Its key parts are:

- *media*—which includes newspapers, magazines, radio, television and advertising;
- *telecommunications*—which includes telephony, data and other information-related products and services; and
- *information technology*—which includes computer-based systems and software.

There are many definitional differences and disputes within this field. The boundaries of these formerly disparate sectors have blurred in recent years as a result of the increasing merging or convergence of these three sub-sets. In some studies, two of the three sectors named above are given a particular focus. For instance, *information technology* and *telecommunications* are often integrated in the literature as *IT&T*, indicative of the strongest of all integration across the three sub-sets. The term *ICT* is also used to abbreviate a category of *information and communications technology* goods and services. Also, with the recent growth of the Internet, the convergence of *information technology, telecommunications and Internet technologies* as *IT&TI* has also emerged. The traditional media industries no longer remain isolated but are increasingly becoming involved with innovative services which involve information technology and telecommunications.

The term 'communications' will be used as the most useful umbrella term to encompass the study of an extraordinary range of activities and relationships, including traditional media program makers, information content providers, telecommunications carriers and associated service providers, information technologists, institutional managers, policy-makers and regulators, and, not least, citizens and users. These are the principal stakeholders of this study.

The book is generally organised chronologically, although several chapters inject a short historical perspective to properly set contemporary issues about a particular subject in context. Chapter 1 deals with the institutional framework of the 'old media', the origins of privileged access to public media assets, how power has historically been concentrated in so few hands, and whether it matters that Australia has an abnormally high level of concentration of media ownership and control. Chapter 2 explores the forces at work driving the changes that have led to unprecedented growth in the field of communications, and the relationship between new communications technologies and globalisation. Catalysts for change involve a complex interplay between technological innovation, convergence challenging the old boundaries, and new forces in the international political economy, as well as the growing marketplace dependence on new communications services. Chapter 3 deals with how the established Australian media institutions are responding to these changes and wondering about their identity in an age of new media, both private sector media conglomerates and public broadcasters, especially the Australian Broadcasting Corporation.

The radical changes in Australian telecommunications resulting from the introduction of competition, and a plethora of new players, are analysed in Chapter 4, as well as the privatisation of Australia's record market capitalisation company, Telstra. Chapter 5 examines the most significant innovation in communications of the past decade—the coming of the Internet—and canvasses issues about its governance, pressures for censorship, virtual communities, changing notions of place and space, and whether a new economy is emerging around the

Internet. The social dilemmas and paradoxes faced in this age of technological abundance are canvassed in Chapter 6. These involve the changing personal and institutional identity associated with communications, the relationship between the 'old media' and the 'new media', cultural responses to globalisation, and equity issues related to the growing disparity between the information rich and the information poor. Chapters 7 and 8 look at issues in the development of an information society, at select examples of what other countries are doing in communications policy, and at the development of the many new on-line services and a related appropriate new communications infrastructure in Australia.

The book has a strong futures orientation in probing how Australia is placed in this more globalised world, and how individually and collectively we situate ourselves within this new international communications order for the future. It offers a look at the new debates concerning the notion of a 'third way', popularised by US President Bill Clinton and British Labour Prime Minister Tony Blair, and systematically relates these to future Australian communications policy. The final chapter argues that Australia must re-think its present communications strategies. A ten-point blueprint is offered in searching for a more imaginative and responsive long-term national communications policy for Australia.

Many of the shifts in communications today are underpinned by technological change on an unprecedented scale. We ought to remember, however, that technology merely enables, and it is actually culture that drives change. In the past half-century we have been clever at inventing several major communications technologies, but we ought to use these technologies in the future to reinvent better and more satisfied selves.

At a time of such complex change in communications, and excessive preoccupation with mountains of instant up-to-date information, there still remains a place for mature reflection about the way things are, and how they might be. This is a book—to borrow a phrase from the Canadian scholar Marshall McLuhan—that sets out to *probe*, rather than to prove.

TREVOR BARR
AUGUST 1999

1. Media moguls: Power, ownership and influence

So there it was—a media 'policy' founded on notions of mates and enemies, just like the third world. If you wanted to succeed as a media mogul, you had to be on side with Bob and Paul. And these men knew what they were doing. In 1987, they allowed the three commercial networks to be sold to buyers who had all borrowed heavily. Soon, the three networks were in receivership at the same time. This is still believed to be a world first.

Les Carlyon, senior columnist, the *Sunday Age*, former editor of the *Age*, and former editor-in-chief, the *Herald and Weekly Times*, quoted from the *Sunday Age*, 11 May 1997, p. 17

Already we can see the process beginning in Australia in the claims by media owners that their businesses would be more efficient and competitive if unhindered by rules providing for diverse ownership through the separation of various forms of media. It is not economic efficiency that the existing rules most impede, however, but massive profits and more centralised media power.

Paul Keating, former Treasurer and Prime Minister, 'The media sharks should be told: No more', *Age*, 19 June 1999, p. 15

Media theory is intellectually situated within political, economic and cultural theory, which classically has been trying to understand how the world works and how it *ought* to work. New debates about new media in an information society raise challenging issues about how the world *might* work. Central to any of these major debates are profound questions about where power resides in society, who has that power, and how they choose to exercise it.

Approaches to understanding society through the study of political economy usually focus on the creation, accumulation and distribution of wealth. Australia's media environment has long been dominated by successful family-owned newspapers which, with the advent of new electronic media, radio and television, diversified their operations to build wealthy and powerful media conglomerates.

Newspaper businesses became media conglomerates over decades of expansionism. The key groups were Fairfax, News Corporation, Herald and Weekly Times (until 1987) and Packer.

Australia's oldest surviving newspaper was founded in 1831 as the *Sydney Herald*, and has been published since 1941 as a regular daily newspaper, the *Sydney Morning Herald*, under the proprietorship of the Fairfax family. Melbourne's *Age* was in the hands of the Syme family from soon after its foundation in 1854 until 1948, when a public company was formed in which Fairfax had a controlling interest. Fairfax later fully acquired the remaining Syme shares in 1983.

Rupert Murdoch began his management career with Adelaide's evening paper, the *News*, in 1956 and soon embarked upon a course of major acquisitions of established newspapers, initially in other cities of Australia but later in the United States and Great Britain, including the *Times* and the *Sun* in London. In 1964 Murdoch created Australia's only national daily newspaper, the *Australian*, which allegedly ran at a loss for over twenty years. Rupert Murdoch, Australia's most remarkable entrepreneur has now built one of the world's largest global media conglomerates, News Corporation.

Melbourne's *Herald*, founded in 1902, was the first newspaper in what was for decades Australia's largest newspaper chain, the Herald and Weekly Times group. From the 1940s to the 1970s, around 40 per cent of all newspapers sold in Australia were published by the Herald group whose most dominant management figure was Sir Keith Murdoch, father of Rupert. In 1987 the Herald and Weekly Times group, with a significant small shareholder base, was acquired by Rupert Murdoch and became part of News Corporation.

The Packer family has long been influential in the Sydney newspaper scene. In 1923 Frank Packer became director of the *Daily Guardian* and *Smith's Weekly*, papers run by his father. In 1933 he launched what eventually became the country's best-selling magazine, the *Australian Women's Weekly*. In 1969 he bought one of Australia's greatest cultural icons, the *Bulletin*, which had been founded in 1880. Frank Packer strategically sold both his daily newspapers, the *Daily Telegraph* and the *Sunday Telegraph*, in 1972 and left one of his sons, Kerry, to focus the Packer group's strategy around commercial television and a stable of magazines.

There has been an increasing tendency towards press oligopoly in Australia. In 1903 the 21 capital city daily newspapers were owned by seventeen independent owners; by 1960 the fourteen daily newspapers had seven owners; and by 1999, two groups owned ten of the twelve dailies in Australia.[1] When radio licences became available in the 1920s and 1930s, it was largely the successful newspaper company quartet of Murdoch, Packer, Fairfax and the Herald and Weekly Times group who acquired them. This pattern of concentrated ownership and control was again perpetuated in the 1950s and 1960s, when predominantly the same groups were successful in gaining most of the lucrative metropolitan television licences, leaving a legacy of strong cross-media ownership. In the mid-1970s in Melbourne, for instance, the Herald and Weekly Times group published Australia's highest-circulation morning and evening newspapers, the *Sun News Pictorial* and the *Herald* respectively, and controlled HSV-7 television and radio station 3DB in Melbourne. This was the typical pattern of strong cross-media ownership exercised by the big four across Australia until the late 1980s.

Three groups still dominated Australia's media ownership and control in the late 1990s: Rupert Murdoch's News Corporation, Kerry Packer's Publishing and Broadcasting Ltd (PBL) (of which Australian Consolidated Press is a wholly owned subsidiary) and John Fairfax Holdings Ltd. The bulk of the ownership of Australian daily newspapers now resides with only two companies, News Corporation and Fairfax Ltd, with only the *West Australian* and *Canberra Times* remaining under separate ownership. Australia's largest magazine publishing group, Packer's Australian Consolidated Press, also controls the country's most successful commercial television network, the Nine Network. With the introduction of pay television in Australia in the early 1990s, News Corporation also became a 50 per cent owner of Foxtel, Australia's leading pay television company. (Telstra was the other 50 per cent owner.) PBL took over half of the News Corporation interest in late 1998. For metropolitan radio, the ownership pattern is essentially duopolistic: the two largest players are Austereo, in which the major interests are held by Village Roadshow, a large Australian film distribution, exhibition and production group, and the Australian Radio Network, which is owned equally by two media groups, Australian Provincial Newspapers and Clear Channel Communications. Australia's pattern of media ownership and control has thus long been essentially one of power residing in the hands of a few well-established corporations, with highly interlinking patterns of ownership and interests.

This integration of press, radio and television ownership and control has given Australia one of the highest levels of concentration of media ownership in the democratic world. Table 1.1 summarises the principal assets in Australia of these three media groups.

Table 1.1 Australia's media ownership and control 1999: Institutional summary

News Corporation (Murdoch holds about 30% of its voting stock)

Newspapers	*Other*
Australian	Festival Records
Herald Sun	Mushroom Records (50%)
Sunday Herald Sun	National Rugby League (50%)
Daily Telegraph	
Sunday Telegraph	
Advertiser	
Courier-Mail	
Mercury	
Northern Territory News	

Subscription television

25% ownership of Foxtel Pay TV

Publishing and Broadcasting Ltd (The Packer family holds a controlling shareholding of 35%)

Australian Consolidated Press (wholly owned subsidiary of PBL)	*Subscription television*
	25% ownership of Foxtel Pay TV
Television	*Other*
Nine Network	Crown Casino
	Hoyts Cinemas

Magazines

Bulletin
Australian Women's Weekly
Woman's Day
Cleo
Cosmopolitan
Dolly
People
Australian Personal Computer
Australian Home and Garden
Wheels

John Fairfax Holdings Ltd

Newspapers	*Magazines*
Age	Business Review Weekly
Sunday Age	Personal Investment
Sydney Morning Herald	Australian Geographic
Australian Financial Review	Shares

Elites: Concentration of media ownership and control

What is the origin of this pattern of Australian media ownership and control? In framing the Federation of Australia in 1901, our founding fathers bestowed great responsibilities on the Australian Parliament for national communications policy. Most of them would turn in their graves if they could see the outcome at the end of the twentieth century. Section 51(v) of the Constitution gave responsibility for 'postal, telegraphic, telephonic and other like services' to the Commonwealth Parliament, and this section has been held since to include broadcasting as well. Print media was not envisaged in 1901 as being subject to Commonwealth laws, although subsequently the structure of newspapers has been affected by other regulations relating to broadcasting, foreign ownership and competition.

There have been several key justifications behind the regulation of Australian broadcasting. The limited electromagnetic space has been a major factor which underpinned government regulation of broadcasting. Licences to broadcast were granted by the Commonwealth government to particular companies, individuals or interest groups. Hence, privileged access was granted to public assets, although licences were intended to carry with them reciprocal obligations. We shall see later in this chapter how the granting of broadcasting licences, and the conditions attached to the holding of such licences, has been the subject of blatant partisan favouritism on the part of the Lyons, Menzies, Fraser, Hawke and Keating administrations during the past 60 years.

Governments also assume that broadcasting has a uniquely powerful impact on society, especially the alleged intrusiveness into people's lives of commercial television and its influential agenda-setting capacity. The political process has striven for an elusive 'balance' in content, usually sought to be administered through a broadcasting authority (the Australian Broadcasting Control Board from 1949 to 1976, the Australian Broadcasting Tribunal from 1976 to 1992, and the less overtly regulatory Australian Broadcasting Authority since 1992). The principal piece of broadcasting legislation in the second half of the twentieth century—the *Broadcast and Television Act 1942*—long held broadcasters responsible for providing 'adequate and comprehensive' programs, although regulators struggled to interpret what this meant or how it might be implemented.

Does it matter, and does anyone care, that Australia's pattern of media ownership and control is so highly concentrated? The arguments against high levels of concentration of ownership may be summarised as:

- potential abuse of power;
- loss of diversity of expression;
- conflict of interest; and
- repressive journalistic culture.

Potential abuse of power

Arguments against concentration of ownership rest on the premise that the unchecked concentration of power is undesirable. Media conglomerates are generally controlled by a small group of people who are unelected and, in many ways, unaccountable to the community at large. Where proprietors have a high level of ownership of a media group, such as the Murdoch family, which owns about one-third of News Corporation, or the Packer family, which also owns about one-third of the shares of Publishing and Broadcasting Ltd, the chances are clearly greater for proprietors to have a dominant influence over editorial policy and content. On the assumption that media are influential in people's lives, it is the arbitrary power, and the *potential* for its abuse, which is of central concern. This is a position consistently taken by Royal Commissions in the United Kingdom and Canada, and was also the finding of a major Victorian government inquiry into problems associated with high levels of concentration of daily newspaper ownership which reported in 1981.[2]

Loss of diversity of expression

In the healthiest democracies, a wide range of diverse opinions are offered, and media offer a plethora of different positions, values and biases. No citizen is obliged to accept any particular position or argument, but the principle of a person's democratic right to put their view is of paramount importance. The most desirable media systems ensure that 'a hundred flowers of opinion' can bloom. The core of the problem with concentration of ownership is *not* that we have a few media demon owners who dream up evil schemes to foist upon the public through their media outlets, or that they can easily manipulate an unsuspecting public, but that we lack sufficient *ideological diversity* within our media system. A key theme of this book is the potential that exists for greater ideological diversity with the new media.

Australian media are essentially dominated by a commercial ideology, and while this is the institutional reality of our economic system, it inevitably restricts the full range of diverse and antagonistic views that the system can offer its media publics. We will not solve the alleged problems of concentration of media ownership by simply adding more players to the system if they are *all* ideological players of the same kind. Take the following theoretical and somewhat fanciful exercise in attempting to change the players who own and control the daily newspaper interests. For the sake of this exercise, picture the Australian capital cities in the mid-1970s, with their dailies owned by the Herald and Weekly Times group's morning *Sun News Pictorial* and the evening *Herald*, Fairfax's *Sydney Morning Herald* and *Australian Financial Review*, and Murdoch's *Australian*.

Just suppose that only one of these commercial papers could somehow remain with the same commercial owner, but that *all* of the others were sold individually to, say, the Australian Council of Trade Unions, the Australian Conservation Foundation, the Brotherhood of St Lawrence and the Australian Council for Social Services (assuming for this exercise that these groups could afford and wanted to buy them!). This would produce an ideological mix within the ownership, where more diverse opinions and alternatives would most likely be offered. When media executives pontificate that we enjoy the world's best television, or boast that Australia has the freest and most open media system, they rarely acknowledge the limitations that the dominance of commercial ideology imposes on the achievement of full diversity. Of course, commercial media are central to our framework of institutional communications, but surely they ought not to dominate the whole system?

Conflict of interest

Media outlets cover the breadth of financial dealings in the Australian business scene, and conflicts of interest can arise where media owners are directly involved with other business activities which inevitably get media coverage in the outlets they control. Vic Carroll, a former editor-in-chief of the *Sydney Morning Herald* and a former managing editor of the *Australian*, made this comment on the problem:

> [A]ll media are potentially compromised when the media owner/controller has substantial non-media interests. They are politically compromised when the non-media interests involve government licences. We do not expect News Corporation newspapers to be vigorously critical of the two airline policy whilst News controls Ansett. We do not expect them to investigate fearlessly the terms and conditions of Fox Studios' deal with

the NSW government over the Sydney Showground. We do not expect the Nine Network's current affairs programs to be profoundly sceptical of casino values. But we should expect someone to be doing these things and that will only happen while there is a diversity and separation of ownership.[3]

It is a contention of this book that we need as many media players as possible to be involved, portraying different perceptions of the world and what it could be like.

Another area of concern about conflicts of interest arises where a Minister with portfolio responsibilities for media policy becomes, or is perceived to become, captured by particular media commercial interests. A good example of this was provided by Peter Westerway, former Chair of the Australian Broadcasting Tribunal, who said at a 1997 conference on media ownership and control:

> [O]ne of Paul Keating's first, if regrettable, acts as Prime Minister was to appoint his old friend, Graham Richardson, as the Minister for Transport and Communications. He became somewhat better known as 'the Minister for mates,' or alternatively 'the Minister for Channel Nine.' Richardson's twisting and turning over policy issues is a story in itself (ably described in Marian Wilkinson's book *The Fixer*). But on the central issue with which we are concerned today—media ownership and control—Mr Packer's warm friend and well paid future employee was very single minded. Take it from one who canvassed no less than four ministers and a prime minister seeking support, that Graham Richardson's mission as a Minister of the crown was clearly the same as it is now: to arrange matters to suit his mates. And no one else in the government—including the Prime Minister—was going to stop him.[4]

Repressive journalistic culture

Significant differences of opinion occur, of course, across the commercial media and within individual media organisations. It is common editorial practice of many daily newspapers to commission senior journalists with diametrically opposed views to write detailed feature articles on major issues. Some newspapers recruit leading journalists, such as Kenneth Davidson or Paddy McGuinness, because of their particular political position, be it left or right wing, in an attempt to contribute to the balance of the paper. There are, however, few examples of the recruitment of highly radical journalists, from any side of the political spectrum, who fundamentally challenge political and social orthodoxy.

Many journalists try to debunk the fears expressed about concentration of media ownership by stating that in all of their years of journalism, no owner has ever told them what to write. Commercial media institutions, however, generate internal pressures which are a product of the character of their ownership. The issue here is *not* that some bosses actually direct their editors or journalists about what they may publish or write, but rather that media personnel *internalise* the values of their organisation and become conditioned by their occupational environment into conventions of particular commercial institutional uniformity. Jock Given, Director of the Communications Law Centre, has alluded to the 'two views of reality' that were offered by the Murdoch and Packer camps about the future of Rugby League in New South Wales during the nasty and protracted negotiations over television rights in 1997. What emerged in the media coverage, he argued, was two images of Rugby League:

> . . . the honest, decent sport set upon by avaricious and unprincipled demons of News Corporation, and a new world of super players, super skills and super entertainment . . . The Murdoch papers have delivered, relentlessly, for Super League. Channel Nine's football coverage, at least until it was supplemented by rights to some Super League fixtures, delivered for the Australian Rugby League (ARL).[5]

This example has been chosen not merely to show that a predictable commercial bias intruded into the media coverage by the two media companies involved in the fight for television rights, but also to demonstrate the pressures towards self-censorship on journalists working within those organisations. How, for instance, could a Packer employee possibly do a pro Super League journalistic piece in this circumstance, or a Murdoch employee write a pro ARL status quo rights story?

Moreover, where media ownership is highly concentrated, the covert pressures to toe the company line are greater because of the lack of alternative employment opportunities. Mungo MacCallum, a flamboyant Canberra political correspondent for the *Nation Review,* observed many years ago that 'you get to know what you cannot write'. Similarly, Max Walsh, an experienced managing editor of the *Financial Review* and the *Bulletin,* once said that 'there are no parachutes for editors'. This pressure for conformity has become more serious in recent years in Australia's smaller capital cities as a result of the abnormally high levels of concentrated ownership of daily newspapers, together with the disappearance of most evening daily newspapers. Where do the journalists who 'rock the boat' unacceptably now find alternative similar employment in the 'company towns' with one daily newspaper owner, such as in Perth, Adelaide, Brisbane and Hobart? One would assume that a business environment with vigorous press competition, plus many alternative avenues for editorial and journalistic employment, would be much more likely to produce a dynamic, diverse and rich content environment.

Broadcasting licences: Partisan patronage

The history of Australia's pattern of media ownership and control shows extraordinary political favouritism and pragmatism on the part of governments towards media corporations judged to best serve the party's interests. Broadcasting policy has been one of the most blatantly politicised and incompetently managed areas of government policy since Federation, by both sides of Australia's major political spectrum.

Australia's media policy history is littered with special manifestations of 'Ozzie mateship' towards media barons by governments of the day. During the past six decades the most notable examples are:

- the Lyons government in the 1930s;
- the Menzies government in the 1950s;
- the Fraser government in the 1970s; and
- the Hawke–Keating administrations (1983–96).

The Lyons government in the 1930s

Rupert Murdoch's father, Sir Keith Murdoch, enjoyed a close relationship with Prime Minister Joseph Lyons when he was at the helm of the Herald and Weekly Times Ltd in the 1930s. Historian Robin Walker has documented an example when Murdoch was rewarded for curbing press criticism of the Prime Minister:

> In October 1934 Lyons wrote to the editor of the *Melbourne Herald* to thank him for his great help in the elections. About a year later Murdoch wrote to Lyons objecting to the recent regulations which curbed the extent to which his and other companies could own a chain of wireless stations. Two days later new revised regulations were gazetted . . . The price that Lyons paid for his support was the suspicion that the Prime Minister was at the bidding of the managing director of the Herald and Weekly Times Company Ltd.[6]

Joseph Lyons thus created the policy precedent whereby the Commonwealth government facilitated increased levels of concentration of media ownership.

The Menzies government in the 1950s

The introduction of television to Australia in the 1950s was accompanied by political interference. In 1958 the Australian Broadcasting Control Board (ABCB) had the authority recommend to the government suitable applicants to be granted television licences. When the ABCB, seeking two licensees for television in each of Brisbane and Adelaide, found that the applicants, the Sydney and Melbourne newspaper groups, breached government policy in that television stations were supposed to be controlled by local interests, they declined to recommend any applicant for a licence to the government. They called for new applications for one licence only in each city to be granted to local business, but the government rejected this proposal—and without any public explanation—in October 1958. After Parliament had risen prior to an election, the Postmaster-General announced that the Adelaide NWS-9 licence had been granted to Rupert Murdoch, owner of the Adelaide *News*, the ADS-7 Adelaide licence had gone to the Adelaide *Advertiser*, owned by the Herald and Weekly Times group, Brisbane's QTQ-9 had been granted to Fairfax, and BTQ-7 had gone to Queensland Press, also owned by the Herald group. It was the Menzies government that set in concrete a media ownership pattern which essentially gave the major television licences to the same big press/radio oligopoly, resulting in Australia having levels of concentration of media ownership unprecedented in the Western world.

The Fraser government in the 1970s

Malcolm Fraser's government (1975–83) also provided some extraordinary examples of political intervention into broadcasting licence allocation, and commissioned some changes to the *Broadcasting and Television Act* which became widely referred to as 'the Murdoch amendments'. Surely it is wrong for governments to rewrite national legislation to suit the interests of a particular company at times which are convenient to individual corporate interests. So, how did this happen?

In the late 1970s Murdoch acquired Channel 10 in both Sydney and Melbourne, Australia's largest city television markets. Murdoch by then had begun his expansionist internationalisation of News Corporation, including the purchase of major American newspaper assets. Since it appeared likely that he had become an American citizen to enable him to purchase such assets, he was cross-examined (by Gareth Evans) at the Australian Broadcasting Tribunal (ABT) inquiry into this acquisition and asked whether he was legally an Australian resident, which the law required of broadcast licensees. This issue became clouded in technicalities about the meaning of citizenship and Murdoch won. Given the extent of his vast international media empire, and hence the limited time spent in Australia, he could hardly be called an Australian resident or Australian citizen. Later he became owner of more than 60 per cent of Australia's major daily newspapers. Moreover, few countries in the world have levels of foreign media ownership anywhere near as high as Australia has allowed.

At the time of the ABT inquiry into the Murdoch takeover, there was a 'two station rule' for television, intended to put a check on further increased concentration of ownership of Australian media assets by limiting the control of television licences to a maximum of two for any one proprietor. Murdoch, who wanted a Melbourne–Sydney television axis, therefore needed to divest a television station licence in Brisbane, which he had held since 1958, or be in breach of the Act. Occasionally, existing licence-holding duopolists bought a shareholding in further licences but later divested these interests in order to stay within the legal limit. There were considerable legal procedural matters involved at the ABT hearing into Murdoch's purchases about 'warehousing' shares during the acquisition. The ABT somewhat courageously ruled against Murdoch's acquisition on public interest grounds, arguing that his control of a Sydney–Melbourne network would unduly concentrate media power. But later an Administrative Appeals Tribunal review overturned the ABT's decision in favour of Murdoch.

Astonishingly, however, as a result of this saga, the Fraser government changed the legislation with what came to be known as 'the Murdoch amendments' which removed the important ABT's public interest powers that previously could be taken into account when deciding on the suitability of corporate licensing changes. The revised legislation redefined the criteria for licensing by omitting concentration of media control in metropolitan areas as an appropriate criterion for consideration. So, as a result of these government policy changes, no longer could the government's regulatory body take into account public interest issues in the allocation of public licences! The sad irony of this episode was that the Australian Broadcasting Tribunal was itself a creation of the Fraser government (in 1976), and the Prime Minister, Malcolm Fraser, had said at the time of its inauguration that its role would be to 'depoliticise' the licensing process! A decade later, Fraser appeared as a private citizen, with another former prime minister, Gough Whitlam, to speak at a public rally in Melbourne's Treasury

Gardens where he vigorously denounced the disastrous media ownership legislative changes made by the Hawke administrations.

The Hawke–Keating administrations (1983–96)

The late 1980s witnessed the greatest spate of takeovers in Australia's corporate history, with extraordinarily bizarre buying and selling of media assets. Though there were some complex policy factors at work, especially the inevitable structural industry changes resulting from the introduction of Australia's domestic satellite system, the fact remains that new government media legislation introduced by the Hawke administration was the catalyst for this spate of media takeovers. In November 1986, amendments to media ownership laws, which changed the numerical limits to television station ownership, facilitated a successful takeover early in 1987 of the Herald and Weekly Times group by Rupert Murdoch. It also created a situation where virtually all metropolitan television licences changed hands, with the result that most of the new owners subsequently ran their newly acquired television networks at substantial losses. Flawed legislation, coupled with one of the most deplorable exercises in Australian corporate sector incompetence, meant that Australia's commercial television industry, whose history really had been 'a licence to print money', went broke for a while. Staggering commercial television industry losses in the late 1980s saw Frank Lowy selling the Ten Network, Alan Bond being forced to sell the Nine Network (back to Packer), and Christopher Skase quitting Qintex and the Seven Network before fleeing to Spain.

How could a government have possibly had a hand in this debacle? With the introduction of the domestic satellite system in Australia in the early 1980s, the issue arose of how to utilise the satellite's capacity to cover the whole of Australia to justify the existence of the system and help Aussat, the domestic satellite company, pay its way. It was necessary to alter the established balance of interests between metropolitan television interests and cosy regional television monopolies (among the most profitable business enterprises in the country), otherwise the satellite system would have been underutilised, thereby incurring huge losses. A proposed system for regional television operators to take up supplementary licences, whereby the regional operator would offer a second service via a supplementary licence, did not work. The regional operators were generally reluctant to change an industry structure which had provided them with such handsome profit margins, and few took up the offer of supplementary licences. While searching for a solution to the satellite problem, the government was attracted to a suggestion that had been made by the Australian Broadcasting Tribunal in 1984 to change the two station rule. The then Minister of Communications, Michael Duffy, subsequently explained that the two station limit rule

> . . . was extraordinary because it equated holding a licence in Sydney and Melbourne with Mt Isa and Shepparton, which in terms of power was considerably different. The rule did not take into account the different populations served by television licensees. Based on the 1981 Census data a licence in Sydney served 3.3 million people, that is 22% of the population. Mt Isa served 23000 people, that is 0.2% of the population. So Channel Nine, at that stage, had 43% of the population, Channel 10 had 43% of the population, Fairfax 31% and the Herald and Weekly Times 28%. But Mt Isa and Broken Hill had 0.4% of the population, and Ballarat and Shepparton had 3.3%. It was a ridiculous rule.[7]

On 21 November 1986, Duffy took a submission to Cabinet favouring the abandonment of the two station rule, replacing it with a provision which allowed a prescribed interest in any number of television licences, provided the combined population of the area serviced did not exceed 43 per cent of Australia's population. Therefore, it was recommended that the two station rule be changed to a system based on the upper limits of the then television ownership audience reach, bearing in mind that the combined Sydney and Melbourne television markets comprised 43 per cent of Australia's population. This ought to have allowed more licences to be held by one interest, while still keeping in check the levels of concentration of television ownership. The Labor government was very divided on this issue, but eventually opted for a proposal from the then Treasurer, Paul Keating, a disciple of deregulation in all industry areas, for an upper limit of ownership reach of 75 per cent which had to open up the floodgates. Minister Duffy stayed steadfastly to his proposed limit of 43 per cent, but Keating, with Prime Minister Hawke's support, won the day—at 75 per cent. Later, the Senate amended the legislation to allow for a maximum audience reach of 60 per cent.

As a trade-off to government members who opposed this new ownership policy, and who were concerned that the Labor Party platform actually pledged diversity of ownership, new regulation was introduced to limit the levels of cross-media ownership. Essentially, the cross-media law provision under Labor was that a company could not acquire a television station in a market where it owned a daily newspaper with more than 50 per cent circulation in the same market. Keating said that owners had to make choices about their medium, that they had to choose to be only *'princes of print'*, *'queens of the screen'* or '*rajas of radio*'! As Rod Tiffen later put it, '[R]estrictions on cross-media ownership were the one fig leaf for maintaining diversity.'[8] However, the consequences of the changes to the legislation brought substantial activity to the share market, and within weeks of the legislation being passed, Murdoch bought the Herald and Weekly Times group, and a corporate circus of television takeovers began with multiple changes of ownership and control of licences. Although this brought some new owners into the system, these tumultuous changes provided no real diversity than before, and ended in financial disaster for many companies.

These policy issues were among the most complex in Australia's media policy history, but in hindsight it is difficult to see that any national benefits have accrued from this bizarre policy period. The Howard government (1996–) subsequently came to office promising a thorough media review of cross-media ownership, but after a protracted debate in the major press outlets during much of 1997, together with powerful lobbying from vested media interests, the Howard Cabinet finally decided, rather symbolically, to maintain the legislative status quo. It must be acknowledged that this prevented the Packer organisation from the possible acquisition of the Fairfax group as a result of legislative change, which Packer had publicly sought, and which would have further reduced the number of principal media players. In May 1999 all three major media groups declared that they were in favour of the abandonment of the cross-ownership limits.

If history judges governments merely on outcomes, it will be especially unkind to the Hawke–Keating governments in the context of Australia's new map of media ownership and control. This Labor government's track media record during 1983–96 was partially responsible for:

- The break-up of the Herald and Weekly Times group, which reduced the number of media oligolopists from four (Murdoch, Packer, Fairfax, Herald) down to three. Paradoxically for a Labor government, the Herald group had the most diverse share register of all the media companies—an example of 'people's capitalism'.
- The greatest spate of takeovers in Australia's commercial television history, with unbelievable and unsustainable prices paid for licences. There were many commercially stupid decisions, but they were facilitated by Labor's new legislation on television ownership and control.
- A commercial television system which focused more programming power on the six Sydney and Melbourne channels of 7, 9 and 10 (and with Sydney more dominant managerially than Melbourne) in

- a television industry now increasingly operated on a national networked basis, paying lip service to regional needs.
- The highest levels of concentration of ownership of daily newspapers in Australia's history, with Murdoch owning more than 60 per cent of the major daily newspapers.

What conclusions can be drawn about the effective functioning of media policy in Australia from these examples of blatant politicisation over many decades? Governments and political parties are dependent on an unelected group in society—the media owners and controllers—for their electoral prospects. A political party elected to form a government and administer the portfolio responsible for national media policies and practices behaves in ways which suggest it feels beholden to keep media members 'on side', otherwise it may risk defeat at the next election. This is one of the most serious and unfortunate conflicts of interest inherent in our system of government. Clearly, this is an area where the democratic process has not worked in the way it was intended to work.

The old media hegemony

The complex issues surrounding the alleged power and influence of media institutions, of who owns and controls media, and how they exercise that power, have always been subject to intense scrutiny by interests vested and otherwise.

Media remain central to most people's lives: the old adage still remains true for many people today, that next to sleep and work, our next most time-consuming activity is attending to media. We use media to construct our version of what the world is like, and what we regard as important issues in society depends in part on how the media choose to represent them.

Every political party pays extraordinary attention to the agenda constructed by the media, and how their particular party interests and performance are portrayed, for better or worse. Media institutions are also central to a great deal of financial activities, including an unprecedented level of media share dealings during the past decade and a half. Hence, media institutions and practices have a special prominence in our society; they are significant financial organisations in our economy, major political influences in the democratic process, and central players in the construction of our sense of place in society.

Some media scholars have drawn upon the economic determinism school of thought proposed by Karl Marx in the 1860s, which has recently come to be revisited in debates about power in our contemporary information society. Briefly, Marx saw capitalist society as being driven by the dual dynamism of investment and profit. For Marx, the central axis of the capitalist society was the connection between the 'mode of production' and the 'relations of production'. The 'mode' constituted the products of a society, whether grown, manufactured or extracted, to be used or consumed. He argued that the 'mode' of production separated capital, which accumulated profits, from labour, the 'relations' of production. Capital would inevitably drive down the costs of labour as low as possible. In Marx's form of capitalism there were only two classes, the owner and the wage slave—each locked in an inevitable class war. Technological development emerged through the logic of capitalistic necessity. For Marx, the only way that labour could 'win' over capitalism was by socialising the means of production through a revolution.

Marxism's economic determinism was encompassed in related debates about class and ideology by an Italian journalist and political activist, Antonio Gramsci, who was a founding member of the Italian Communist Party in the 1920s. He was imprisoned in 1926 where he wrote a major treatise before his death in 1937. The notion of *hegemony*, essentially defined as *the power or dominance that one social group holds over another*, is attributed to Gramsci. He was opposed to Marx's strict economic determinism, and preferred interpretations which centred on the role of the human agency in understanding class and other social struggles in historical change. Gramsci gave attention to the structuring of authority and dependence on symbolical environments that correspond to the ruling economic class in a society. According to Gramsci's theory of ideological hegemony, mass media are tools that ruling elites use to 'perpetuate their power, wealth and status [by popularising] their own philosophy, culture and morality'.[9] For Marx, the centrepiece was the economic class struggle, whereas for Gramsci it was more critical to understand *consciousness*, especially how the ruling class had the power to frame or define reality in a way that perpetuated their class domination.

For much of the twentieth century, Australia had among the highest level of concentration of media ownership and control in the democratic world—Australia's media ruling class. Although all were conglomerates, there were significant differences in terms of the content offered, their political outlook and their style. Their management invested in each new medium as it arrived, were enormously competitive, suspicious of each other, and made good profits as media became a significant growth industry of post-Second World War Australia. Each of the major technological shifts was spaced with decades of institutional adjustment, notably the advent of radio in the 1920s and 1930s, television in the 1950s and 1960s, and the introduction of an Australian domestic satellite in the 1980s.

For much of the twentieth century, Australian media expansionism was built around 'press moguls'—the Fairfaxes, the Murdochs, the Packers—as well as the Herald and Weekly Times group (until 1987). Since the 1960s about 90 per cent of what Australians read in newspapers, hear on radio and watch on television are products of these companies. Their collective ownership and control of so many media outlets has given them a commanding political base. James Lull argues,

> [O]wners and managers of media industries can produce and reproduce the content, inflections, and tones of ideas favourable to them far more easily than any other social groups because they manage key socialising institutions, thereby guaranteeing that their points of view are constantly and attractively cast into the public arena.[10]

It is important to point out that there are no uniform social responses to the content offered by major media to their media publics. Inevitably, there is an extraordinarily different set of human responses by media consumers to media outputs. Meaning always remains in the eye of the beholder. It is in this sense that the term 'mass media' has become discredited. Those who own and control the media systems cannot *manipulate* the reading, viewing and listening publics, although they *do* possess the capacity to set agendas of discourse and exert influence in ways that others cannot. Owners can still make the ultimate decision on what is published, and what is not. Editors, journalists and program makers inevitably *construct* the content of media.

Some of the rhetoric of new media suggests that the old power bases are now being eroded by new players and the many new modes of communication, so that the concerns about highly concentrated levels of ownership and control are no longer well founded. With so many new choices along the superhighway, why worry about the old media conglomerates? We are witnessing, so the argument goes, the 'demassification' of the new media.

However, we need to be reminded that the key institutional media power bases still remain firmly entrenched in terms of their market share of audiences. The major audience trends regarding the time devoted to media sources by Australians may be summarised as follows:

- Overall Australian media consumption by adults has not changed greatly during the past ten years. Mainstream media, including radio and television, tended to maintain their number of users, though there has been some decline in hours spent with particular media.[11]

- A.C. Nielsen research in 1977 argued that the average television viewing time was 3.2 hours per day, with 4.4 hours per day for those aged 55-plus, but people in pay television households split their viewing between an average of 2.1 hours daily on pay television, and 1 hour a day on network television. By 1999 Nielsen research indicated that pay TV viewing accounted for 46 per cent of all television viewing in homes with a pay TV connection.[12]
- The number of hours spent over a year (1997) watching commercial television dropped to about 15.5 billion, down by about 470 million, with the difference explained in terms of greater time devoted to the Internet. Heavy users of the Internet tend to devote little time to commercial radio and television.[13]
- There has been a substantial drop in newspaper *circulation* in recent years: total average daily sales have fallen 38 per cent in the past 25 years, and 32 per cent for the period 1986–97, although weekend and Sunday newspapers have fared better. In 1963 one person in three bought a daily newspaper, but now only one paper is sold for every 7.5 people.[14]
- Meanwhile, the amount of newspaper *reading* appeared not to be in decline, with total hours increasing by 120 million in 1998.[15]

Classic Marxist/Gramsci theorists would point out that some fundamental characteristics of Australian media still remain with us at the beginning of this new century—a highly concentrated pattern of commercial ownership and control by the 'ruling class', with most Australians high consumers of the products of the ideological hegemony, although with some significant changes in media habits beginning to emerge.

The era of the twentieth century media age was essentially built around press dynasties: Keith handed on to Rupert Murdoch, Frank to Kerry Packer, and James to Warwick Fairfax. It will be fascinating to see whether the new generation of daughters and sons, notably Elisabeth and Lachlan Murdoch and James Packer, will be able to maintain the dynasties' power base in the age of the new media. Their new century will be commercially and strategically much tougher. The communications technological landscape is in the process of changing rapidly. There are no governments to lobby to acquire Internet advantage for their companies because there are no Internet licences in Australia.

There are radical forces for change under way, to which we must now turn.

2 Forces for change: Communications as catalyst

A technological revolution, centered around information technologies, is reshaping, at accelerated pace, the material basis of society. Economies throughout the world have become globally interdependent, introducing a new form of relationship between economy, state and society, in a system of variable geometry.

Manuel Castells, *The Rise of the Network Society*,
Blackwells, Oxford, 1996, p. 1

A global culture also knows no period, no past, no sequence nor determinate process. It has no beginning, no development, no goal. It is here and now and everywhere . . . A global culture is artificial.

Anthony Smith, 'Is there a global culture?',
Intermedia, vol. 20, no. 5, 1992, p. 11

Every society is an information society. Throughout history, different cultures have adopted different modes of communication, but all are information societies of some kind. Traditional cultures were predominantly dependent on forms of interpersonal communication. The invention of printing led to the penny press, and gave mankind a print culture, possibly the most critical factor in the rise of the nation state. The mid-twentieth century witnessed the emergence of institutionalised forms of major media, notably radio, film and television, often misleadingly referred to as 'mass media'. The labels 'mass media' and 'mass communication' are both contradictions in terms, given the extraordinary disparities of access to the new communications technologies and services within different nations. Moreover, responses that individuals, audiences and consumers make to media content, across different cultures, can hardly be adequately categorised as 'mass culture'. The products of the convergence of communications have also meant that contemporary information society is highly dependent on information networks that can distribute images, data and symbols.

The conventional way in which we have come to use the term 'information society' today is usually dependent upon the degree to which new communications technologies have infiltrated a culture. Rarely challenged is the presumed nexus between 'an advanced information society' and the technologies of convergence—a sophisticated telecommunications system, the widespread adoption of computing and the acceptance of enhanced media systems. There is a popular notion of a historical hierarchy for societal evolution: from 'traditional' societies of hunters–gatherers, to 'least developed' agrarian societies, to 'developing societies' built around agriculture and mining, to 'industrial societies' dependent on smoke stack manufacturing, to the final evolution of an 'advanced post-industrial society'. This final stage is widely believed to depend centrally on the level of advancement of the information revolution. What distinguishes the information society from its predecessor, the industrial society, is that nineteenth-century industrialisation was dependent on energy production and associated manufacturing, whereas the information society is centred on knowledge and information networks, products and services.

Rarely questioned is the proposition that the development of a high-technology society is the best option for all cultures. The literature of the school of 'development communication', notably from Asian and African societies, challenges conventional technological development assumptions by advocating a thorough investigation of the most appropriate technological choices for particular cultures, based on an examination of different options suited to different communications needs, rather than simply accepting every new communications technology that comes along.

Why is there now such extraordinary growth and dependence on information systems and processes? Where does this unprecedented attention and energy, focused around the world's communications industries, come from? What is behind these changes, and who is driving them? In seeking to find answers to these questions, we might explore several major forces at work in the formation of a new convergent communications era, notably globalisation and information technology and telecommunications networks (IT&T), the associated re-evaluation of the role of the nation state, and the drive for privatisation and deregulation.

Technological processes: Convergence and digitalisation

The information revolution is in part driven by major technological changes in communications. The past decade has seen extraordinary technological innovation and application in the world of communications. The key technological changes are summarised below.

Convergence

Traditionally the various media and information forms were reasonably distinct: television channels received and broadcast television programs, telecommunications enabled phone calls to be made between persons, and computing essentially processed information. These formerly discrete functions have now tended to come together, and *the boundaries between media, telecommunications and computing have blurred in a process called convergence*. One can now sit at a personal *computer* to send an e-mail *message* that is transmitted through a *telecommunications* system to a friend overseas. Or, one can watch a horse race on television and soon afterwards see the dividends on screen (*media*), delivered through a Telstra landline (*telecommunications*) from a central dividend database in the TAB (*computing*). This process of convergence is a major force for change in the growth of the new media.

Broadcasting and telecommunications have historically been distinct, both as processes of communication and in terms of the institutions responsible for those services. Television was a broadcast medium, run by television networks, or by ABC-TV, and Telecom, the telephone company, carried voice calls, so essentially content (*media*) and carriage (*telephony*) were institutionally distinct. However, with the introduction of pay television, new kinds of companies emerged: Foxtel was created as a merger between Telstra, which had the telecommunications network, and News Corporation, which had the media programs. Media companies have also created mergers with computing companies, such as alliances between Packer's PBL Ltd and their broadcasting interests in Australia, and Bill Gates' computing software giant, Microsoft, as ninemsn. Technological convergence has also meant greater integration of computing with telecommunications, notably with the Internet which needs Internet service providers (ISPs), of which there are about 800 in Australia, to hook up users to a computer and a modem to get them into the Internet via the international telecommunications network of Telstra.

Figures 2.1 and 2.2 show the key emergent trends of convergence. The three main sectors of the communications industry—media, infor-

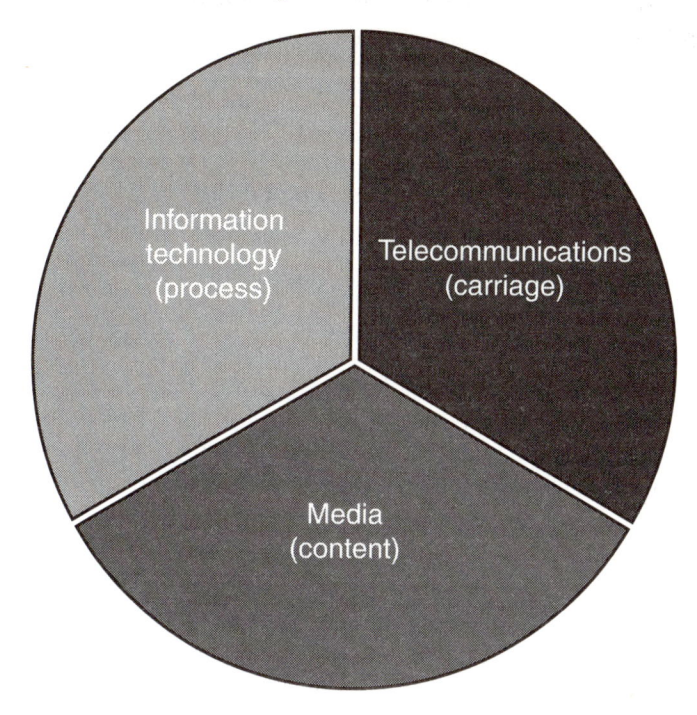

Figure 2.1 Functional convergence

[Pie chart showing three equal sections: Information technology (process), Telecommunications (carriage), Media (content)]

mation technology and telecommunications—have each emerged from different institutional backgrounds, with different patterns of ownership and control. The overall institutional mix of public and private ownership and control in these sectors tends to be largely comparable in Western developed nations, including Australia. Traditionally, print media have had a strong privately owned daily press, publishing and magazine sector. Broadcasting ownership has exemplified the mixed economy, with strong commercial television and radio networks, and significant public broadcasting organisations, notably the Australian Broadcasting Corporation (ABC) and Special Broadcasting Services (SBS). The information technology sector, however, has been dominated by transnational ownership of entrepreneurially brilliant international computer corporations, such as IBM, Hewlett Packard, Compaq, DEC, NEC and Apple, who have their good times and bad times, mostly good, but who come and go as members of this most maverick sector of this industry.

The ownership and control of the telecommunications sector has essentially been different again. Traditionally part of the public service utilities, telecommunications carriers tended to have originated as government departments of post and telegraphs that were generally

Figure 2.2 Institutional convergence

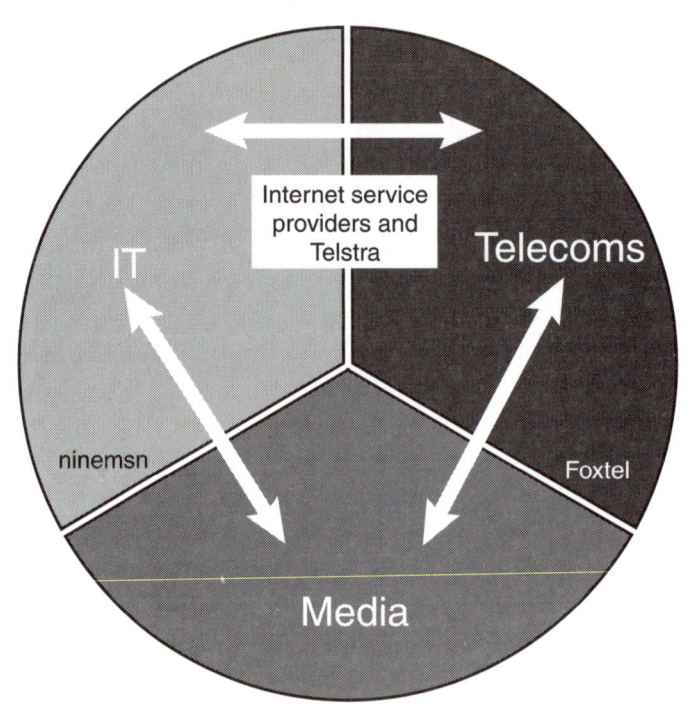

[Diagram: Circle divided into three sectors — IT, Telecoms, Media — with bidirectional arrows between them. Labels: "Internet service providers and Telstra" between IT and Telecoms; "ninemsn" between IT and Media; "Foxtel" between Telecoms and Media.]

separated functionally in the 1970s and 1980s into posts and telecoms. Telecommunications carriers have become a favoured target for private investors since their rapid growth in the mid-1980s. The tensions of purpose endemic in a shift from public to private ownership have led to a fundamental review of the *raison d'être* of telecommunications. Convergence is not only blurring the demarcation lines of services, but is also leading to a questioning of the fundamental nature of the established communications institutions. Figure 2.3 shows the way industry boundaries will be much less clear in the future.

We are likely to see a burgeoning array of new media services in the next few years, and there is no end to speculation about the possible future of these convergent services. For example:

- Operating systems will interconnect television sets, mobile phones and household appliances. Sony sees the home of the future with hundreds of connected intelligent devices and consumer electronic appliances, all 'seamlessly connected'.[1]
- Three companies already have mobile phones that are Web enabled. British Telecom Cellnet predicts that more than 40 per cent of Internet traffic will come via a mobile phone within three years.[2]

Figure 2.3 Convergence of industries

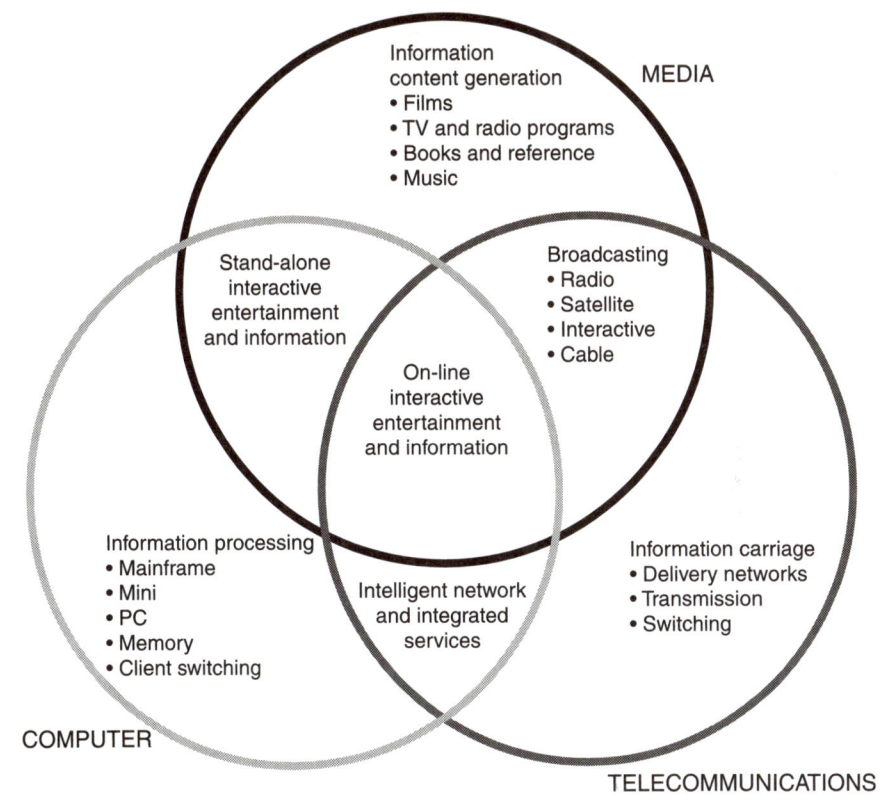

```
                    Information
                    content generation    MEDIA
                    • Films
                    • TV and radio programs
                    • Books and reference
                    • Music

        Stand-alone              Broadcasting
        interactive              • Radio
        entertainment            • Satellite
        and information          • Interactive
                                 • Cable
                    On-line
                    interactive
                    entertainment
                    and information

  Information processing              Information carriage
  • Mainframe                          • Delivery networks
  • Mini         Intelligent network   • Transmission
  • PC           and integrated        • Switching
  • Memory       services
  • Client switching

COMPUTER
                                       TELECOMMUNICATIONS
```

Source: CIRCIT.

- MP3 players will enable users to download music from the Internet into a portable unit eventually capable of storing 300 hours of audio.
- New watches will double as cameras, with a port allowing them to connect to a video recorder or TV.
- Electronic books will have on-line capacity to download purchases of new titles.

Future new media services are likely to see more examples of integration between traditional stand-alone devices, such as television sets, and the newer networked appliances and services. There are endless 'what if' predictions.

Convergent alliances

The increasing tendency for media, telecommunications and information technology to merge functionally, and thereby to need hig capacity networks to deliver new services, has been a driving force behind recent institutional changes. Telephone carriers, broadcast and film companies, and computer companies have looked closely in recent years at the possible benefits of forming mergers or alliances with organisations that would complement and strengthen their core business. Frank Blount, when Managing Director of Telstra, said:

> [T]he dawning of the information age not only signals the end of an era in the telecommunications business, but the start of a different business paradigm, where the rigid boundaries between industries break down, where previously distinct functional categories become blurred, and allow the resulting cross-pollination to generate new forms of business creativity.[3]

Convergence has generated greater cross-pollination of advertising and marketing agreements. In January 1999 Murdoch's Fox television network and the Internet company Yahoo! bartered a deal on the cost of advertising on the Fox television network for Yahoo! in return for the promotion of Fox sitcoms in on-line Net chat sessions. In February 1999 Murdoch's News Corporation and Packer's PBL announced they were forming an alliance to take a $429 million investment in a newcomer telecommunications company in Australia called One.Tel. The alliance included a deal of $30 million a year in free advertising across the Packer and Murdoch worldwide media interests to promote One.Tel as a global telephone company. These kinds of interlinking advertising and marketing agreements will continue to flourish in the future to the great benefit of the major corporations.

At another level, telecommunications companies have increasingly become involved in greater examination of their ownership base and many have embarked upon major acquisitions and mergers of unprecedented financial magnitude. Collectively they actually represent the largest financial transactions in the history of corporate capitalism. The intense competition for global markets has also driven the industry towards greater amalgamation. The recent major mergers and alliances are briefly outlined in Table 2.1.

Some huge alliances proposed between cable television companies in the United States and telephone companies were called off during the early 1990s, just before their marriage, with the prospective partners sensing that they would be unable to work closely together, settle their differences and achieve acceptable mutual profit returns. Failures include proposed mergers or acquisitions between IBM/MCI, AT&T/NCR, MCI/BT and members of the Concert group. However, the degree of successful major mergers has led to heightened speculation that the telecommunications industry is becoming so dominated by the

Table 2.1 Recent major telecommunications mergers and alliances

- British Telecom (BT) and America's AT&T announced in July 1998 a US$10 billion joint venture to link their international businesses to serve US, European and Asian markets. They subsequently took a 30 per cent share in Japan Telecom.

- The AT&T–BT merger followed BT's failed US$24 billion bid to acquire US rival MCI WorldCom Inc's acquisition of MCI Communications Corp., America's second-largest long-distance telephone carrier, was one of the largest merger deals in history at US$37 billion.

- In June 1999 AT&T, America's largest long-distance telephone company bid US$62.5 billion for Media One group, the number four cable television provider, topping a US$56.5 billion bid from Comcast.

- In June 1999 SBC Communications and Ameritech merged at a cost of US$62 billion to create a unique US 'national-local' focus US telecoms company.

- In June 1999 Deutsche Telkom offered US$95 billion—Europe's largest ever merger—for Telecom Italia, which was fighting off a takeover bid by Olivetti. The new group would be capitalised at 173 billion euros and have some 72 million fixed-line customers.

Sources: Company data; *Australian Financial Review*; Paul Budde Communications.

world's super carriers that eventually there will be only a few gigantic global communications corporations. The ownership of the world's telecommunications business is now concentrated around several huge corporations, notably AT&T–BT, MCI WorldCom, SBC–Ameritech and Deutsche Telkom–Telecom Italia. Each of these corporations is likely to embark upon further aggressive mergers and acquisitions throughout the world in the next few years.

Institutional convergence brings together different corporate cultures. Considering the complexity and global dimensions of some of these alliances, it is little wonder that there are substantial problems to be overcome. Concert, which had promoted itself as 'the first communications company for the world', struggled to integrate the different corporate cultures of BT and MCI and was reported to have lost between US$125 billion and US$175 billion a year.[4] Where telecommunications carriers and television networks come together in a partnership, there appears to be some corporate cultural incompatibility. This has been experienced with the creation of Foxtel, a subscription television company formed between an alliance of Telstra, the biggest Australian network operator, and News Corporation, one of the world's largest publishing companies. The telecommunications carriers have tended to be quasi-bureaucratic, technology-focused and driven by an engineering ethos. The publishing culture, on the other hand, has tended to be more financially free-wheeling, with higher risk-taking, but more customer-focused. Similarly, ninemsn (PBL/Microsoft) has faced creative tensions within its organisation because the television people tend to want to build web sites to promote the network's television programs, while the Microsoft people want to use the web sites to shift the television audience to on-line purchasing.

Considering the complexity and global dimensions of some of these alliances, it is little wonder that there are substantial problems to overcome. The formation of an alliance can mean loss of managerial autonomy, technology sharing with the possible further loss of core competence to a partner, and difficulties in reconciling global and national strategic objectives. The trend towards alliances is overwhelmingly defensive in nature. The prime paradox of convergence is that while it has led to greater *institutional consolidation*, the process offers a greater *diversity of services* than ever before. Corporations are desperately trying to work out the way these new divergent markets will go in terms of consumer choices, and to position themselves for any unforeseen major market shifts.

Digitalisation

Telecommunications networks traditionally used analogue technologies in the supply of their core product, plain old telephone services, but they have progressively converted to digital technologies. Digitalisation is the process which converts any type of information into a compressed form to be sent as a stream of bits for use at the receiving end. Digitalisation enables the transmission of all kinds of communication signals—not only voice, but also data, video, graphics and music—over the network. In the past two decades or so, almost every telecommunications system in the world has been expanded and modernised to some extent through digitalisation. Digitalisation has facilitated the rapid traffic increases of multiple modes of communication.

Computing processing power is doubling every eighteen months or so, and the technical capacity to transmit information over networks is currently growing by about 20 per cent a year. In computing, microprocessors have progressively become more powerful, smaller and more affordable. In a sense, convergence and digitalisation have enabled telecommunications systems to become gigantic computer networks with unprecedented capacity to distribute extraordinary quantities of information in many forms. The electronic superhighway is really a term used in 'media speak' to refer to a series of interlinked information networks which allow individuals and organisations around the world to exchange information, sometimes in vast quantities, as text, audio or video.

Increased delivery capability

The communications revolution is an age of technological abundance in terms of the additional ways in which different media forms and information can now be delivered. Television services have been enhanced by the 30 or so additional channels that can be offered by pay television operators, such as Foxtel or Optus Vision. The video cassette recorder has enabled some viewers to buy or rent programs, especially their favourite movies, and also provides greater viewing flexibility to record and replay television programs as and when they wish. Telephones have been reinvented, notably as mobile phones which enable callers to call the person rather than the fixed handset. To cope with the exponential increase in traffic, telecommunications companies have had to upgrade their networks and invest in new high-capacity fibre optic networks, such as those that now exist between every capital city in Australia. A mainframe computer that used to take up a whole room in a warehouse can now sit on the desk as a personal computer and provide more capacity and adaptability than its predecessor. A small laptop computer can offer word processing, a facsimile service and access to the Internet.

Previously constraining technical and financial factors, such as the limited bandwidth space for television or, for so long, the high costs of rolling out high-capacity intelligent terrestrial telephony networks, have been changing in recent years. This is not to say that the new intelligent information networks come cheaply, but that higher-capacity networks can now be built with greater comparative affordability to the carrier because of the range of new value-added services that can be marketed. The capability of satellite technology is improving substantially, with larger, high-powered, longer-life satellites being launched into low, medium and geostationary orbits. There are some extraordinarily ambitious international satellite ventures, some of which are likely to struggle for a long time to recoup their massive capital outlays. If there is any aspect of contemporary information society that is clearly revolutionary it is the technical capability, reach and intelligence of information networks. Ours is an age of abundant technological capacity; the most vexed issues surround how we develop the services we can now transmit more readily, and how we explore what users want and are willing or able to pay for.

Network intelligence and interactivity

Convergence in itself has not involved the invention or introduction of a new technology; rather, it involves the way communication technologies that have existed for some time have come together to facilitate wider, more integrated methods for the distribution of information. Digitalisation is the compression information engine driving changes within these new high-capacity networks of distribution that convergence has made possible. These factors made possible another critical technological change which has ushered in the age of new media—network intelligence and interactivity.

The advanced telecommunications network is now sometimes referred to as an 'intelligent network'. The old telephony system basically offered one-to-one voice communication. In a sense those networks which carried the telephone calls were 'dumb' in that they simply sent telephone calls down the 'pipe'. The new networks not only have greater capacity to carry more messages, but the revolution in software has enabled new functions to be added. In making a telephone call to an organisation today, one may be automatically digitally switched through to the particular person to whom you wish to speak, or the call may be queued, or one may be put on hold while another call is taken, or one may be able to record a message on the person's voicemail. So, with digitalisation and switching, a telephone consumer can have intelligent services added, such as voicemail, call waiting or call line identification.

More widely, a range of new on-line services is also emerging, requiring users to make their responses interactively, which involves more than simply speaking in response to an incoming call. The advanced stages of convergence see the digitalisation of content, such as film and television programs, databases, and print and library materials, delivered over a telecommunications network to households. We now have the prospect of many new on-line services emerging, such as home shopping, home banking, home gambling, video on demand and electronic commerce (e-Commerce). These require consumers to make judgments about whether they want access to particular on-line services, and the process of interactivity may involve some form of processing a payment for the service. Moreover, a variety of interactive multimedia services has recently emerged, with some brilliant creative integration of voice, video, graphics, music and data. Table 2.2 shows the progression of some major services emerging as a result of convergence, digitalisation, capacity, and intelligence and interactivity.

In summary, these technological forces for change mean that we have never had so many choices in terms of the variety of forms and processes of communication, which collectively represent a technological revolution in itself. Nations are constructing globally interconnected, high-capacity, high-speed information networks to accommodate the

Table 2.2 Migration trends to new media/communications services

	Existing	Extensions
Modes of communication	Phone Fax E-mail Internet	Videophone Video mail Teleconferencing Electronic public meetings
Domestic media	Radio Television Newspapers Computer games	Video on demand Electronic newspapers Virtual reality Digital radio
Information/Data	Audio text Teletext Yellow Pages Videotext	On-line directories Database searches Interactive advertising Electronic newspapers
Transactions	Telephone banking Credit card bookings Mail shop catalogues EFTPOS (Electronic funds transfer at point of sale)	Home shopping Home banking Home gambling Tele-referenda (voting from home)

Note: These are trends. The line between existing communications services and extensions is not intended to represent clearly delineated boundaries between them.

increasing demands of users and consumers for new communications services. Technology, however, is never a neutral force. The way human beings use these communications technologies, and other technologies, is increasingly reconstituting our lives in ways we are trying to understand. Just as the technological forces driving change are now global, we must now turn to the complex human dimensions of global change in this new phase of our communications history.

Communications in the global marketplace

Globalisation is one of the most vexed and encompassing forces affecting people today. It is a term subject to definitional disputes, although clearly we are witnessing the reorganisation of the world's regional, national and international economies, and trading practices on a global scale. While we may not yet have an integrated planetary economy, where one currency is used worldwide, many economic, social and cultural processes are becoming more globalised. John Tomlinson has suggested that 'globalisation may be most broadly described as a process in which complex interconnections are rapidly developing between societies, institutions, cultures, collectives and individuals world wide'.[5]

A new technological revolution, based on information technologies, has enabled and facilitated the financial interdependence of economies around the world. The dramatic fluctuations of world currencies and stock markets in October 1987 and October 1997 demonstrated the world's neurotic financial dependence on the vagaries of international capital markets. In late 1997 and early 1998, Asian economies, especially Indonesia, South Korea and Thailand, faced the extraordinary turmoil and pain of dramatic currency devaluations on world financial markets and associated huge write-downs of their local share values. Capital is now managed around the world in globally integrated financial markets which enable, for the first time in history, billion dollar transactions to be carried out in seconds through electronic information circuits. The growth of capital transactions on foreign exchange markets has been astonishing in recent years. As Kelvin Rowley points out, in 1970 almost all currency trading was done in the London Eurodollar market and the annual turnover was estimated at US$59 billion.[6] However, by 1990 the average volume of foreign trading stood at *US$600 billion per day*,[7] and by 1998 foreign exchange transactions had risen to *US$1.5 trillion per day!* This represents about 60 times the volume of world trade in goods and services.[8] The efficiency and ubiquity of information technology networks has made possible such extraordinary shifts of financial information around the world's capital markets.

In his seminal 1996 work, *The Rise of the Network Society*, Manuel Castells wrote:

> [I]n this electronically operated global casino, specific capitals boom or bust, settling the fate of corporations, household savings, national currencies, and regional economies. The net result sums to zero: the losers pay for the winners. But who are the winners and the losers changes by the year, the month, the day, the second, and permeates down to the world of firms, jobs, salaries, taxes and public services.[9]

Hence, because currencies have become increasingly interdependent, so too have national economies. The world has developed much more complex global interconnections, undeniably in terms of the operation of international capital markets, facilitated by globally interconnected computer networks which can immediately process and interchange financial data in every stock market in the world. There is a school of thought which argues that a new *global information-based economy* is emerging. Castells postulates that the present world economy 'is global because the core activities of production, consumption, and circulation, as well as their components (capital, labor, raw materials, management, information, technology, markets) are organised on a global scale'. And it is *informational*, he adds, 'because the productivity and competitiveness of units or agents in this economy (be it firms, regions or nations) fundamentally depend upon their capacity to generate, process and apply efficiently, knowledge based information'.[10]

This new global economic order has been tooled by innovation in electronics and information software which has made possible the rise and reach of contemporary network societies. The Castells thesis proposes a fundamental nexus between the development of a global informational economy and IT&T networks.

For Castells, a new world is taking shape at the end of the millennium. Globalisation has created a place where our planet is organised around a common set of economic rules, and this ought to bring new opportunities to those nations who can trade widely and well. But it may also mean that it has become virtually impossible for any nation to remain economically autonomous. Since Australia has chosen to be a member nation of global capitalism, and had little choice other than to do so, any attempt to run a fortress Australia national economic policy could lead to dire consequences in terms of our currency and asset valuation, with crippling social penalties. So, how much economic autonomy do we still have?

Much of the literature on globalisation implies that such changes are inevitable in a world where apparently all that mere mortal human beings can do is adapt to the 'borderless world' as best they can. Resistance to the onslaught of global economic forces remains marginal. Dr Mahathir Mohamad, Prime Minister of Malaysia, said in 1997, 'We are told that we must open up, that trade and commerce must be totally free. Free for whom? For rogue speculators? For anarchists wanting to destroy weak countries in their crusade for open societies.' Globalisation, unlike gravity, is not really uncontrollable, as Les Macdonald has argued:

> If we accept, even for a moment, that the course of events within, for example, the international trade markets are subject to conscious human decision and intervention, it requires action from our political leaders to control and direct those markets, to ensure they produce better outcomes for all citizens of the world. Just as the world trade system and the world financial system are constructs which were formed out of concerted international action we, as free nations, have the capacity to again change and to recognise error.[11]

Yet how can nations act against such powerful forces for change? In the wake of the Asian and South-East Asian international financial crisis post-late 1997, interventionist measures have increasingly been speculated upon, and considered by particular governments, to redress the economic vulnerability of the impacts of globalisation. Currency boards, which fix a country's exchange rate at an unchangeable level against a hard foreign currency, are one such measure. In early 1999 there were fourteen countries which had currency boards. Another idea is the notion of creating a new exchange rate structure for all countries based on the US dollar and the gold standard. Obviously, this would be difficult to achieve for all countries. There have also been proposals to create stronger regional monetary blocs, notably for Europe and Asia or South-East Asia, which would also need a common level of agreement among member nations to be effective. Also proposed has been the introduction of taxes on foreign exchange transactions in order to slow the international movements of speculative funds, especially to reduce dramatic flights of capital when there is a run on a particular nation's currency. All such measures, however, remain peripheral in terms of the way global capitalism now operates.

Despite widespread concern with the havoc that the 'global casino' can inflict upon particular economies, and the range of possible alternative remedial actions that have been floated, the forces for globalisation appear to be gathering momentum. Most governments and corporations appear to have accepted that these forces for change are unstoppable, and that the best strategy is to find ways to capitalise on the opportunities that come with globalisation. Australia has always traded well internationally, and overseas sales of its agricultural products, energy products and raw materials made us one of the most prosperous countries in the world during the 1950s and 1960s. But internationalisation has become globalisation—and information has become the commodity of the future. As the *Australian Financial Review* editorialised,

> The end of the Cold War (the counterpart to the 1815 defeat of Napoleon) has greatly expanded the size of the global market economy, from Russia to China to Brazil. And the information revolution (akin to the industrial revolution) is radically cutting the costs of communication and increasing the returns in knowledge.[12]

It is imperative that Australia finds its place in this new global knowledge society.

Globalisation, the nation state and communications

Globalisation has brought into question the role of the nation state in contemporary society. For most of the twentieth century the nation state, (for example, France, or Indonesia, or Australia) with its political, legal and administrative control over territory with boundaries, was the predominant form of governance of people throughout the world. Anthony Giddens defines the nation state, which exists in a complex of other nations, as 'a set of institutional forms of governance, maintaining an administrative monopoly over a territory with demarcated boundaries (borders), its rule being sanctioned by law, and direct control of the means of internal and external violence'.[13] So, what happens when governments, which only have sovereignty over their own territory, find that their borders are breaking down? Castells observes that 'state control over space and time is increasingly bypassed by global flows of capital, goods, services, technology, communication and information'.[14] With globalisation, argues Castells, the nation state 'is losing its power, though importantly, not all of its influence'.[15]

The rules within which economic activity takes place are increasingly being defined in international fora such as the World Trade Organization (WTO), the International Monetary Fund (IMF), the Organisation for Economic Co-operation and Development (OECD), the World Bank, and the Group of Nations, G7. Regional trading blocs are influential, too, notably the Asia-Pacific Economic Co-operation (APEC), the Association of South-East Asian Nations (ASEAN) and the North American Free Trade Association (NAFTA).

Nation states, it seems, are now trapped by the global directives and assessment of their overall economic policies. Global movements of capital, goods and services, technologies and information increasingly bypass state boundaries and thereby undermine state control. Importantly, though, nation states can still make influential economic policy decisions, such as the privatisation of national assets, deregulation and competition policy, and local industry policy. The communication system designed by nation states—their press, publishing, broadcasting and telecommunications institutions, and how these have been allowed to operate within the nation—have often been a central factor in the maintenance of state power. However, this mode of control, too, is now increasingly being challenged by the content and form of information which can be made available to citizens in a globally interconnected network society. For Manuel Castells, 'with this eventual defeat will come the loss of a cornerstone of state power'.[16] Or, as Anthony Smith has argued, the new information society will expose the actions of governments more so than ever before, making them more accountable, and less able to keep their domestic media systems closed and controlled.[17]

The world's political order has radically altered during the past decade: communism has collapsed and capitalism has gone global. Who would ever have believed that a President of Russia, Boris Yeltsin, would be elected and re-elected in the 1990s? And who would ever have believed that a Chinese President, Jiang Zemin, in September 1997, would decree that China will move progressively into a mixed economy where private ownership would become the norm, and where the right to vote will grow incrementally from its present level of Chinese rural shire councils? Castells referred to the 'collapse of statism', pointing out that even the few exceptions, such as North Korea, Cuba and Vietnam, are in the process of linking up with global capitalism. He added that the new global capitalist system, linked by 'its relatively successful *perestroyka* and the emergence of informationalisation' has become the technological basis of world economic activity and social organisation.[18] As a result of these dramatic shifts in international political economy, the world's largest communist nations are in the process of radically changing their communications industries.

Markets for products have become more universal. For decades, the nation state was the dominant political framework for the management of production in a world where capitalist competition was basically territorial, and where industrial competition operated among nations. Domestic companies grew into large corporations by successfully exporting their products. Increasingly, they *internationalised* (not globalised) their businesses by initially marketing goods and services to other countries from their base in their nation state, then later internationalised their operations.

Now, the respective components of production may be dispersed far from the corporation's geographic base, with a variety of inputs from different parts of the world. For example, until 1997 a Swedish-based telecommunications company, Ericsson, manufactured telephone exchanges in Australia using electronic components imported from Malaysia and printed circuit boards from Singapore. It assembled the exchanges in Melbourne, installed them for Telecom Australia, and later won substantial telephone exchange export business throughout Asia from Melbourne. This was the nature of the value chain in international manufacturing in IT&T. The tendency in the late 1990s is how to concentrate communications equipment manufacturing around fewer and fewer plants, and to draw upon those international suppliers who offer the most cost-effective prices, irrespective of their location.

For Alain Benoist, world markets have become increasingly independent of a particular nation or territory: '[T]erritory is being replaced by network, and economic and political space are no longer bound together in a deeper meaning of globalisation.'[19] The industrial locus of power has shifted to the world's transnational corporations which have experienced extraordinary growth in the past 30 years. Benoist offers this summary of their growth, power and reach:

> Multinational companies are those that do more than half of their business abroad. In 1970, there were 7000 of them. Today (in 1996) there were 40,000 and they control 206,000 subsidiaries whilst employing only 3% of the world's population (about 73 million). The budget of these corporations in 1991 was greater than all of the world's exports of goods and services ($4.8 trillion); directly or indirectly they control a good third of the world's revenue and the top 200 of these companies monopolise a quarter of the world's economic activity. Nearly 33% of world trade now takes place among the subsidiaries of the same corporations, not between different corporations. These network corporations have immense resources at their disposal.[20]

The ethos of the transnational corporation is to strive continually for greater productivity and competitiveness in order to maximise their profitability. Many governments of nation states now vigorously court transnational corporations and offer a whole series of incentives for them to become involved with their national strategic development plans and operations. Deals include reduced local taxation, particularly favourable import or export agreements, lower tariffs on gas and electricity, rent-free premises at technology parks, and so on. These corporations naturally play governments off against each other to get the best incentive deals in return for some kind of technological development agreement with a particular host nation state. Singapore, for example, with its three million people, centred its information technology development strategy on attracting the technological expertise of multinational corporations. As a result of this strategy, Singapore is a technological client state of particular multinational corporations and consequently remains industrially vulnerable, but the policy has delivered considerable prosperity to this naturally poorly resourced island state. By 1995 the per capita income of Singapore exceeded that of France.[21]

Capitalism has undergone major restructuring during the past two decades. Rising levels of national debt in the 1980s contributed greatly to monetarism and economic rationalism becoming the predominant paradigm of capitalism's economic strategy—another major trend of globalisation in the past twenty years. United States President Ronald Reagan's 1980s' slogan, 'best government is the least', encapsulated the prevailing mood of most Western democracies as they undid the welfare state, reformed industrial relations and attacked labour movements. The fundamental role and purpose of government itself has been brought into question. A key part of the traditional role of government was to define citizenship, impose law, protect public health, provide general education and maintain security, with a wide range of public services paid for through taxation. However, with rising public (and private) debt levels in most economies, trading in an increasingly competitive globalised world, a fundamental shift of economic priorities occurred towards more corporate, privatised, deregulated, competitive and liberalised domestic industrial policies, and a reordering of established welfare-based social policies.

This economic policy shift is generally referred to as *economic rationalism*, neatly interpreted by Brian Toohey:

> [E]conomic rationalism was based on the premise that markets know best. Rather than being seen as integral to its operation, governments are considered as somehow outside—and antagonistic to—the economy. Accordingly, governments should not 'intervene' in the market. As a corollary, individuals should be required to do more to fend for themselves, taxes should always be cut, privatisation should be accelerated, communal services funded according to the 'user pays' principle and deregulation of all markets, including the labour market, pursued with unremitting vigour.[22]

For an economic rationalist the processes of globalisation provide commendable ways of testing any economy's level of economic efficiency, and only the fittest deserve to prosper. In advocating economic systems based on the unrestrained forces of free market competition, they generally argue that vulnerable economies are those which have inefficient government practices and intervention in the free market.

Since the late 1970s many economies have embarked upon strategies that dramatically shifted away from state-owned and controlled enterprises towards a privatised, deregulated and more competitive economy. This was perceived to be important not merely in terms of comprehensive reform of the domestic economy, but also in response to the pressures driving towards greater internationalisation. Around the world, many state-owned monopolistic telecommunications carriers have been progressively privatised, wholly or partially, beginning in Britain in 1984, then Japan's NTT in 1986–87, Hong Kong and Chile in 1988, and Malaysia, Argentina and New Zealand in 1990–91 (see Table 2.3). The international telecommunications privatisation program was estimated to have raised about US$104 billion between 1981 and 1991, with estimates of revenue made in 1992 for the following decade (1992–2002) of privatisation then expected to raise about US$350 billion.[23] This is likely to be a gross underestimate of the total privatisation capital raising for that decade.

Most nations are now positioning themselves to find their place within the global economy. Most nations are now making major

Table 2.3 Major privatisation of telecommunications carriers/operators, 1984–99

Australia	33.3% of Telstra sold in 1997, plus another 16.6% in 1999. Government policy to eventually sell 100%.
Canada	All operators privately owned.
France	20% of France Telecom sold in 1997.
Germany	26% of Deutsche Telkom sold in 1996.
Japan	Partial privatisation since 1986, but minimum 33.3% public ownership of Type 1 carriers.
Italy	Telecom Italia fully privatised by 1997.
Mexico	Telmex partially privatised in several tranches between 1990 and 1994.
Netherlands	PTT Telecom fully privatised.
New Zealand	Telecom New Zealand fully privatised in 1990.
Sweden	Telia fully state-owned.
United Kingdom	British Telecom fully privatised.

Sources: *World Telecommunications Development Report*, 1994 and 1997–98; and *Telstra Privatisation: Background Report*, CEPU, June 1998.

strategic planning decisions to develop their communications sector to ensure long-term prosperity. Most are now striving to achieve the economic determinism of an information economy. The world's major communications corporations are trawling the globe for a bigger slice of the action. The forces driving the information revolution are gathering momentum, although the extent of societal change engendered by the information economy is unevenly experienced globally, within nations and by their citizens. Australia must build its own best version of a knowledge society within the new global society. We turn now to look at how Australia's major media institutional players are responding to these forces for change.

3 Challenge of change: Australia's media institutions

> *Whatever happens to Murdoch, he has an importance far beyond himself. What will matter are the choices of the barons who control the fantastic new holdings of the global village . . . No one has elected them to such a responsibility. Technology, the market, and in Murdoch's case, invincible energy and ambition have given it to them.*
>
> William Shawcross, Murdoch, Simon & Schuster, New York, 1993, p. 424

> *The ABC is an invisible net which binds us as a nation . . . Hundreds of voices, each individual, each different, often opposed—but maintaining civil debate, passionate, but tolerant.*
>
> Newsletter, Friends of the ABC (Vic. Inc.), Issue 5, August 1998, p. 12

Gone are the days of a comfortable Australian commercial media environment, delivering somewhat predictable annual financial returns to shareholders. The commercial networks' view of the Australian media environment was that advertising charges could be hiked up regularly, and they were 'supplemented' by one major public broadcaster whose role was to produce 'alternative' programming which could never challenge their market dominance. The 21st century opens, however, with an institutional media marketplace in transition where stability will be constantly challenged by new forms of media. Australian citizens' traditional media habits—essentially based around reading newspapers and magazines, listening to radio and watching television—are in a state of flux.

The 1990s saw the emergence of new media forms to produce a much more complex and fragmented marketplace. So, what is the current scorecard for the established Australian media industry? To what extent have established media habits altered in this new environment?

How have the institutions of communications adjusted to convergence and seized opportunities in a 'borderless world' which is now supposed to be everyone's oyster?

Major media: The commercial club

Rupert Murdoch's News Corporation

Australia has produced some remarkable media entrepreneurs and especially Rupert Murdoch who, although he later chose to become an American citizen, built one of our few home-grown transnational corporations, News Corporation. Rupert Murdoch is arguably the world's most successful commercial practitioner of globalisation in the field of communications. Starting with a newspaper he inherited from his father in 1956, Adelaide's afternoon tabloid, the *News*, Murdoch has built a A$54.5 billion (in assets) global communications conglomerate during the past four decades, based on gross revenues of A$18.9 billion (1988) of which only A$1.85 billion is Australasian. Rupert Murdoch is a highly intuitive business strategist—erratic, a huge risk-taker, an interventionist operationally and politically, ruthless with his senior personnel, but by far the most commercially expansive globalist of all Australian media entrepreneurs. Tables 3.1, 3.2 and 3.3 show the remarkable commercial growth and success of News Corporation.

Murdoch's initial growth came from the expansion of his newspaper business, which is unusual in that this happened at a time when many media proprietors were trying to shift their assets out of daily newspapers and into electronic media, especially television. Murdoch's level of acquisition of newspapers around the world, and his capacity to transform

Table 3.1 News Corporation, 1997 and 1998: What the empire is worth (A$)

	1997 Assets $bn	1997 Revenue $m	1998 Assets $bn	1998 Revenue $m
Newspapers	7.4	3 158	8.2	3 773
Magazines and inserts	3.7	1 541	6.6	2 493
Television	13.4	3 700	17.4	4 865
Film entertainment	4.5	4 219	5.9	5 897
Book publishing	1.8	946	2.3	1 087
Other	10.5	825	14.1	834
Total	**41.3**	**14 389**	**54.5**	**18 949**

Source: News Corporation 1998 Financial Report, p. 21.

Table 3.2 News Corporation, 1997 and 1998: What the empire earns—Operating income (profit after abnormals but before income tax) (A$)

	1997 ($m)	1998 ($m)
Newspapers	610	669
Magazines and inserts	388	586
Television	573	930
Film entertainment	133	375
Book publishing	16	55
Investment income	408	31
Net interest expenses	(634)	(770)
Other	20	31
Operating profit before abnormal items	1433	2068
Abnormal items	(569)	(118)
Operating profit before income tax and outside equity interests	864	1950

Source: News Corporation 1998 Financial Report, p. 21.

Table 3.3 News Corporation, 1997 and 1998: Revenues by geographic areas (A$)

Area	1996 ($m)	1997 ($m)	1998 ($m)
United States	9 056	10 054	14 002
United Kingdom	2 335	2 665	3 095
Australasia	1 697	1 670	1 852
Total	**13 088**	**14 389**	**18 949**

Source: News Corporation 1998 Financial Report, p. 2.

them into profitable companies, far outweighed the rate at which newspaper circulation was shrinking. From his newspaper beginnings in Adelaide in 1956, Murdoch moved to Sydney in 1960 where he acquired regional publisher Cumberland Newspapers, followed by Mirror Newspapers, publishers of the then popular afternoon tabloid, the *Daily Mirror*. Rupert Murdoch has been accused of replacing his love of the smell of ink (developed as a boy wandering around the floor of his father's newspaper, Melbourne's *Herald*) with a love of the smell of money. However, he established the first national daily, the *Australian*, in 1964, which ran at a loss for about the next twenty years. In 1972 he extended his Sydney newspaper base with the purchase of the Sydney *Daily Telegraph* and *Sunday Telegraph* from Sir Frank Packer. In 1979 Murdoch launched an unsuccessful takeover bid for the Herald and Weekly Times group, but his more audacious bid in 1987 was successful, giving News Corporation a stranglehold over daily newspaper ownership in Australia. Murdoch now owns seven of Australia's twelve daily newspapers, and

Table 3.4 Australian newspaper circulation, 1967–98: Net paid circulation ('000s)

Daily newspaper	1967	1976	1999
National			
Australian	84	126	132
Financial Review			91
New South Wales			
Daily Telegraph	340	334	432
Sydney Morning Herald	290	270	233
Daily Mirror	362	389	Closed
Sun	346	342	Closed
Victoria			
Age	186	230	193
Sun	629	628	Merged Sun-Herald
Herald	501	450	554
Queensland			
Courier-Mail	248	269	216
Telegraph	167	167	Closed
South Australia			
Advertiser	207	233	206
News	145	165	Closed
Western Australia			
West Australian	182	242	220
Daily News	96	118	Closed
Tasmania			
Mercury	51	55	50

Source: Audit Bureau of Circulations and company data. The figures as at 30 September for 1967 and 1976, and 31 March 1999 are net paid circulation. The *Australian* and *Sydney Morning Herald* are publishers' estimates for 1967 data.

Note: News Corporation Ltd-owned newspapers in 1999 are in italics.

he has shut down several major papers that were not performing to his commercial satisfaction. Three states—South Australia, Queensland and Tasmania—now have only one locally published daily newspaper, owned by News Corporation, bringing new meaning to the term 'company town'. The dominance of News Corporation in the Australian daily newspaper market today is shown in Table 3.4.

Murdoch's international newspaper push began in the United Kingdom in 1969 when he bought the weekly *News of the World*, followed by the acquisition of a London daily, the *Sun*. Both increased their circulation and earning power under Murdoch's management. The *Sun* became the largest-selling English-language daily newspaper, with sales growing to four million papers a day in the United Kingdom, giving it 50 per cent of the daily tabloid market. His *News of the World* remains the largest-selling Sunday newspaper in the English-speaking world. Murdoch rocked the British establishment when England's premier thunderer, the *Times*, fell to him in 1981. Murdoch then turned his attention to the United States, with notable acquisitions being the *San Antonio Express* (Texas) in 1973, then the *National Star*, followed by the purchase of America's fourth-largest daily newspaper, the *New York Post* in 1976, the *Boston Herald* in 1982 and the *Chicago Sun Times* in 1983. These newspapers have generally been transformed under Murdoch management; usually to become more tabloid, more popularist, more attractive to a wider base of advertisers and more profitable. Here was a man not heeding Marshall McLuhan's prediction that print was a dead medium.

While Rupert Murdoch became the largest owner of mass circulation newspapers in the world in the 1970s and 1980s, his strategic focus in the 1990s has become the development of worldwide multimedia platforms for the future, with colossal investments in unproven high-risk satellite-based new media ventures. Murdoch's global satellite interests include 40 per cent of B Sky B (a British satellite venture that was initially a financial disaster), a quarter share in the Japanese venture, J Sky B, a US$250 million investment in a planned Indian service, I Sky B, a 65 per cent stake in pay TV from Telecom Italia for US$108 million, and a 30 per cent share of a Latin American satellite group, DTH. In 1993 he paid US$825 million for Star TV, a Hong Kong-based satellite operation with its footprints across 53 Asian countries, though it has been unprofitable since its purchase.

In the United States, Murdoch was able to achieve the commercially astonishing feat of building a successful fourth over-the-air television network, Fox Television, to compete with the entrenched financial network powerhouses of ABC, CBS and NBC. In 1994 News Corporation bid a record US$1.58 billion for the four-year rights to the American National Football League, ousting incumbent CBS. In 1997 Murdoch's full-year profits failed to impress the financial markets—so much so that he was forced to try to extricate himself from the potentially disastrous US$3 billion risk for his planned US satellite television network venture, American Sky Broadcasting (A Sky B). He initially attempted to sell its assets to Primestar, but when this was blocked by the US Justice Department, he entered into a joint agreement with Echostar Communications Corporation. Murdoch intuitively mixes his assets of mature growth-oriented established media businesses with his sense of the best new investments in media developments.

In July 1998 Rupert Murdoch announced his intention to spin off the multi-billion dollar American television, film and sports assets—to become the Fox Entertainment group—and initially proposed selling a stake of up to 20 per cent to the public. The Fox group was to be composed of three divisions, the Fox Filmed Entertainment group, the

Fox Television group and Cable Networks. News Corporation would maintain full ownership of its publishing assets in the United States, Britain and Australia, the international pay television and satellite assets, and other interests. Murdoch's intention was to strengthen the balance sheet, reduce the gearing and 'put some pep' into the stock price of his holding company, News Corporation. After selling at a high of US$25.63 per share just after its float in November 1998 (which raised US$2.8 billion from the US market), Fox's stock price fell by about 20 per cent during the next few months. Media analyst Jill Krutick, of Salomon Smith Barney, advised investors to stick with Rupert:

> Fox is an entertainment juggernaut launching into the next millennium with an impressive array of assets primed to take advantage of News Corporation's diversified distribution platforms. Fox boasts an enviable sports franchise, a full TV syndication pipeline, a rich library, lucrative TV stations group and several key ventures nearing a profit.[1]

In April 1999 Murdoch bought the US Liberty media group in a complex script partnership valued at US$2.85 billion to give News Corporation full control of its US cable TV programming and distribution assets, including Fox Sports Network, Fox News Channel and Fox Family Channel. The deal meant that Rupert Murdoch had become the first Australian-born businessman to have acquired A$10 billion in personal wealth.

There are three critical features to note about Rupert Murdoch's corporate empire. First is the extraordinary ambition and level of risk of its management style and process, led by the man himself. Rupert Murdoch has purchased many newspapers during the past 30 years that no one else wanted to buy and they generally became commercially successful under his managerial and journalistic formulae. He is now in the process of attempting to establish an extensive network of high-risk but potentially lucrative electronic media satellite platforms, which News Corporation will own and control in the major regions of the world. In the 1998 News Corporation Annual Report, Rupert Murdoch told shareholders that with the launch of satellite businesses in Latin America and Japan, the company's global reach was close to 75 per cent of the world's population.

Second, his strategy is that of a global capitalist extraordinaire at the corporate level, capable of facilitating a management process from which can emerge a blockbuster investment film such as *Titanic* (which allegedly generated more profit than the rest of the Australian media combined in 1998), of paying US$320 million for an American baseball team, the

LA Dodgers, and bidding (although unsuccessfully) A$1.6 million to take over England's Manchester United Football Club!

Third, his managerial response to convergence has been to build a vertically integrated corporation with substantial newspaper, magazine, book publishing, television network and subscription TV interests, the Twentieth Century Fox film operation, and the emerging capital-intensive distribution platforms of international satellites. For Murdoch some commercial alliances are necessary, but he prefers to stand alone as much as possible.

News Corporation has highly capital-intensive investments in software development, such as Fox Sports, Fox Kids and Fox News, as well as substantial US cable-based distribution assets in development, notably Fox Sports and the Fox Family Channel. However, even Rupert Murdoch is grappling to achieve a credible commercial Internet strategy; he is still floundering after announcing three attempts during the past three years. He is on the record as saying that News Corporation would not spend lavishly on the Internet. For Rupert Murdoch the new media still have a long way to go commercially.

Describing News Corporation Ltd as the best big stock on the share market in mid-1998, Australian financial analyst Alan Kohler pointed out Murdoch's remarkable commercial unorthodoxy:

> [B]y all things institutional investors say they hold sacred, News Corporation should be the worst performer. It has a too powerful chairman–chief executive who practises breath-taking nepotism, its reporting standards are derisory, its structure is too complicated, and its executives make little or no effort to properly explain it to the market.[2]

The most fascinating question about the future of News Corporation is whether Rupert is immortal. Son Lachlan is Senior Executive Vice President of News Corporation; daughter Elisabeth is Managing Director of Sky Networks at British Sky Broadcasting; and his other son, James, is President of News America Digital Publishing. In July 1999 Rupert Murdoch nominated Peter Chernin, formerly Chairman of Twentieth Century Fox's movie operation, as his nominated successor as President and Chief Operating Officer of News Corporation. Long term, Lachlan appears to be the heir apparent when the great commercial man is no longer able to wield his incredible entrepreneurial flair and influence.

Kerry Packer's Publishing and Broadcasting Ltd

The Packer family has long been one of the most dominant commercial and policy forces in Australian media. Kerry Packer's corporate strategy has been different from Rupert Murdoch's in that Packer built his growth on expansion within Australia rather than attempting a comparably ambitious but high-risk international acquisitive strategy. The centrepiece of the Packer commercial strategy has been the cash powerhouse of Australia's most successful long-standing commercial television network, the Channel 9 network, together with a 40 per cent share of the Australian magazine market. The Packer family's multiple attempts to take over one of its major competitors, John Fairfax Holdings Ltd, have failed, primarily due to the continuation of government regulations regarding cross-media ownership, particularly the prevention of joint ownership of a major daily newspaper together with a commercial television channel in the same capital city of Australia. In 1997 a campaign waged by the Packer camp to get the Howard government to abolish the cross-ownership rules narrowly failed, preventing a probable Packer takeover of the Fairfax group. The issue of media cross-ownership surfaced again in mid-1999, but it is most unlikely that either of Australia's major political parties will take the risk of changing media ownership legislation in a way that could lead to another spate of senseless takeovers. Unless the Packer camp become globalists like Murdoch, they may continue to have limited major growth opportunities within Australian communications in the future.

The key Packer company businesses are:

- Nine Network Australia (NNA), the core free-to-air television business;
- Australian Consolidated Press (ACP), the core magazine business; and
- PBL Enterprises, the group's other businesses and investments.

Table 3.5 shows the impressive financial growth of the Packer empire during the years 1993–98, significantly in traditional media, and helped considerably by a revaluation of its television licences in 1997 from $554 million to $1.32 billion, although markets would put

Table 3.5 Publishing and Broadcasting Ltd: Balance sheets, 1993–98 ($'000)

	1993	1994	1995	1996	1997	1998
Total revenue	562 323	610 512	953 001	1 316 462	1 201 532	1 730 936
Income tax	(40 071)	(47 784)	(69 888)	(85 524)	(87 938)	(57 001)
Net profit	**70 363**	**88 300**	**149 937**	**225 387**	**182 116**	**476 443**

Sources: *Australian Financial Review Shareholder*, 13th edn, Ian Huntley Pty Ltd, 1998, p. 411; and Publishing and Broadcasting Ltd Annual Report, 1998, pp. 2, 45, 48.

the value of their television licences at far more than the revised book value.

The Packer family has successfully ridden on the growth of the established media businesses, especially the dramatic growth of television advertising revenues during the 1990s. In television, Australian advertising revenue expenditures have more than doubled during the past decade, rising from $1.1 billion in the year ended June 1987 to $2.4 billion in 1997.[3] This is principally due to the fact that capital city network television in Australia has long remained a regulated commercial oligopoly around Channels 7, 9 and 10, as well as being protected by a major competitor in pay television until 1995 and a moratorium on the introduction of advertising on pay television until July 1997.

Packer has been more successful in television programming rights acquisitions than he has in acquiring other media companies, like Murdoch. Most notable was his bold and eventually highly profitable attempt to take over traditional Test cricket, where he initially set up World Series Cricket in opposition to Test cricket. Eventually the Australian Cricket Board caved in to Packer on Test cricket rights and awarded Channel 9 a lucrative ten-year exclusive television contract. Channel 9 has consistently topped the overall commercial television ratings under Packer management, and Australian Consolidated Press remains Australia's largest magazine producer. PBL shares topped $10 in May 1999 which resulted in Kerry Packer's personal fortune leaping to $6.4 billion, up $1.2 billion on 1998, according to *Business Review Weekly*'s richest 200 list in May 1999. This is a spectacular increase on the Packer personal wealth estimated at a mere $200 million back in 1984. The old media have indeed been kind to Kerry.

Kerry Packer has been one of the most influential figures on Australian media policy in the past twenty years. It was Kerry Packer who in 1979 convinced the Prime Minister, Malcolm Fraser, that Australia should establish its own domestic satellite system. While the government did not follow the Packer policy model of a privatised satellite system, government-owned Aussat emerged as Australia's satellite company which eventually became a catalyst for the introduction of a more competitive telecommunications environment, forced major changes to the rules of media ownership and control, and the break-up of Australia's cosy regional television monopolies. Those close to media policy formulation in Canberra, in both Labor and Coalition governments, have suggested that the persistent Packer policy advocacy behind the scenes has been critical to many media outcomes announced by governments during the past twenty years. Tony Staley, as Liberal Communications Minister, achieved the remarkable feat in 1980 of getting Malcolm Fraser's Cabinet to approve the introduction of cable television into Australia, only to see his successor, Ian Sinclair, reverse the policy following strong lobbying from the television networks. John Button, Labor's Industry Minister for over a decade from 1983, wrote in his memoirs: 'Hawke had a capable and courageous minister in Michael Duffy but his own prime concern seemed to be accommodating the media tycoons Murdoch and Packer. In Cabinet discussions on media issues these two were like Banquo's ghost loitering behind the prime minister's chair.'[4] Bureaucrat Peter Westerway was critical of Prime Minister Keating's appointment of Graham Richardson as Minister of Communications who, he said, became better known as 'the minister for mates' or 'the minister for Channel Nine'.[5]

The big question for the Packer organisation in the operational handover to third generation James Packer, son of Kerry, is whether it can successfully diversify its operations, especially into new media. James Packer is Chairman of PBL and also Managing Director of the Packer private company, Consolidated Press Holdings, with Kerry remaining as Chairman. Kerry Packer obviously remains highly influential within the group, particularly with the diversification strategy into new gambling asset investments. The Packer response to gaming and gambling emerging as Australia's second-largest industry was to pay $342 million to take out the Showboat Inc.'s Management (of Nevada) contract at Sydney Casino, and to take a 10 per cent stake in the Sydney Casino company in early 1997, which supplemented an earlier 9 per cent stakeholding in Crown Casino, Melbourne's casino, which had originally fallen on hard times.

In September 1998 Melbourne's Crown Casino announced a loss of $350 million, after suffering an operating loss of $68.3 million and making several write-offs. Kerry Packer emerged as the financial knight in shining armour to offer a capital injection of $425 million—in return for a half share of all future cash flows—but the move was blocked by institutional investors. In December 1998 he launched his next attempt to buy Australia's largest casino. PBL unveiled a plan to offer Crown shareholders one PBL share for every eleven Crown shares, which revalued Crown at $550–600 million; PBL also offered Crown's biggest shareholder, Hudson Conway, $130 million in cash—an offer too good to refuse. Victoria's casino regulator approved the merger in May 1999, thereby consolidating the $1.8 billion casino group into the $4.8 billion PBL media empire. Crown Executive Chairman, Lloyd Williams, explained that his long-term friend Kerry Packer 'has the Midas touch' when it comes to timing brilliant financial moves.

The Midas touch has been with Kerry Packer for a long time. In

1983 he sold the Channel 9 network to Alan Bond in an extraordinary deal: Bond paid Packer $805 million in cash, $200 million in preference shares in Bond Media, and held options for 50 million more shares in a billion dollar deal. Just before these preference shares fell due for redemption in March 1990, Bond was in dire financial trouble. In July of the same year, Packer used this leverage to buy back from Bond a controlling interest in Channel 9 for $200 million—or in real terms, nothing!

More recently, Packer's dabbling with Optus saw him sitting on a paper profit of $95.4 million on his $119.3 million investment in Cable and Wireless Optus shares after the telecommunications group's successful share market listing. PBL later (April 1999) told the market that it had sold most of its stake for a pre-tax profit of $140.6 million. PBL had also held an option to equalise its pay television interests with News Corporation by 31 October 1998 as part of a peace deal on pay television with News Corporation. Packer increased the PBL shareholding in Foxtel to 25 per cent (the same as News Corporation) despite its estimated $500 million losses to date. Foxtel is the leading Australian pay television operator, but it is unlikely to yield high returns to investors for several years to come. In response to the crazy prices paid on global equity markets for virtually anything to do with the Internet, PBL announced in February 1999 that it was considering a partial float of PBL Online. June 1999 saw the share market launch of Packer's newly created ecorp—a union of its ticketseller, Ticketek, on-line sharebroker Share Trade and 50 per cent of ninemsn. Issued at $1.20 a share, ecorp hit $1.90 on opening day, netting PBL a paper profit of about $295 million. Although a neat return, ecorp's price suffered on opening day from heavy falls for many key Internet stocks on the New York market on the previous night. Overall, PBL had made about $1 billion on its corporate investments in six months.

The positioning within the new media most likely to test James Packer's business acumen is how he develops PBL's Internet ventures in the long term. PBL Online and its joint venture with Microsoft, ninemsn, is an attempt to build a strong Internet brand in Australia. Despite Microsoft's sensational global commercial success with its computer software, its American on-line content ventures, notably its US$425 million investment in WebTV, has produced dreadful financial returns. Web television has not brought with it the comparable assured advertising dollars that come with over-the-air network television. PBL's Online Chairman, Daniel Petre, was quoted in February 1999 as saying that although the company's main asset, ninemsn, was generating 'millions' in revenue from advertising, sponsorship and the resale of content, and was expected to grow by 50–70 per cent next year, they 'did not anticipate profitability for another two years'.[6] Later, in April 1999—a month before the ecorp float—Petre said:

> [A] year ago ninemsn was getting around six to eight impressions a month. Now we are getting more than 100 million impressions a month. We have Hotmail, the largest e-mail service in the nation. A year ago we had just 350,000 accounts. Today we have 1.4 million and we're adding 100,000 a month.[7]

Industry estimates of the accumulated losses of PBL's Internet business ventures ran at $30 million by mid-1999. At least the ecorp opening day stock market performance represented some pay-back to Australia's largest television, magazine and casino group.

The John Fairfax stable

John Fairfax Holdings Ltd is the corporate outcome of the Fairfax family business which began more than 150 years ago with the *Sydney Morning Herald*; by 1996 it had become a billion dollar Australian company. In the late 1980s, members of the Fairfax family—James O. Fairfax, Sir Warwick Fairfax, Sir Vincent Fairfax and John B. Fairfax—controlled more than 50 per cent of the group John Fairfax Holdings Ltd. The history of Fairfax is that of a company which has long maintained a fierce sense of independence in both its journalistic and corporate practices. It has tended to be a company which has strongly focused on running its business around media substance and working on consolidating its assets through better circulation and attracting more advertising, rather than being diverted by the corporate games, which it has not played well.

Fairfax, not naturally disposed to predatory corporate practices, emerged badly bruised in the aftermath of a spate of crazy media takeovers at the end of the 1980s. This period saw the company launch an unsuccessful defensive bid in January 1987 to beat Murdoch for the Herald and Weekly Times group, saw the sale of Channel 7 in Sydney, Melbourne and Brisbane (to Christopher Skase), and saw Warwick Fairfax, the 26-year-old son of Sir Warwick and Lady Fairfax, launch a disastrous bid to buy the company in September 1987 such that nine banks in January 1988 brought legal action against Fairfax for repayment of loans totalling $500 million. Young Warwick's catastrophic commercial failure saw the company change its ownership base away from a family newspaper dynasty to a corporation with a diversified shareholding.

For much of the 1990s Fairfax has focused on trading itself out of financial trouble in the aftermath of media takeovers, remaining an extremely attractive commercial media acquisition and feeling itself constantly under siege. In the mid-1980s the Hawke Labor government, in a decision seen by some of its supporters as ideological treachery, allowed a Canadian, Conrad Black, to buy 25 per cent of Fairfax, a level above the normal Australian foreign investment limit and well in excess of the foreign ownership limits for newspapers that most Western democracies allow. Black, Chairman of the Hollinger group of Canada, later became frustrated with his attempts to further increase his shareholding in Fairfax and, in December 1996, he sold the greater part of Hollinger's shareholding to Brierley Investments Ltd (BIL) of New Zealand. Hollinger investors who originally bought into Fairfax in 1992 saw their shares more than double in price by the time of the bail-out. Many market analysts believed that BIL was a short-term investor waiting to sell out at a profit, just as it did during its highly profitable strategic purchase of Herald and Weekly Times shares just before the Murdoch takeover offer in 1987. The company denied this, but in December 1998 Brierley Investments sold the bulk of its investments in Fairfax for $600 million.[8] The Packer camp has also periodically bought and sold into Fairfax up to its legal investment limit under cross-ownership regulations (or had to sell down early in 1999 when it slightly exceeded the limit), according to their reading of the regulatory winds of change in the hope that one day they could launch a full takeover bid. Unless the cross-ownership provisions are abolished, or Packer is prepared to sell Channel 9 again, which is most unlikely, their 'dream' of owning Fairfax is unlikely to be realised.

The Fairfax group boasts an array of quality newspapers as well as many of the nation's finest journalistic achievements. Its premier newspaper stable—notably the *Age*, *Sydney Morning Herald* and *Australian Financial Review*, collectively remain Australia's most analytical and influential press, making a major contribution to setting the political agenda. Its principal magazine interest is the BRW group, publishers of the only Australian stand-alone business magazine, *Business Review Weekly*, as well as *Shares*, *Trendex* and *Personal Investment*. With the loss of its considerable metropolitan television assets as a result of the hurly-burly of the takeover era in the late 1980s, it is remarkable that Fairfax could achieve a record-level billion dollar revenue base in 1996 from its almost exclusively print media assets. Strong revenue growth has continued, reaching $1.1 billion in 1998 with the annual net profit up 52 per cent to $118.8 million—a remarkable turn-around from its takeover times. In the opinion of some of its most experienced journalists, recent years have been tough within Fairfax in terms of savage

Table 3.6 John Fairfax Holdings Ltd: Balance sheets, 1993–98 ($'000)

	1993	1994	1995	1996	1997	1998
Total revenue	771 354	846 592	948 433	1 006 148	1 027 881	1 153 602
Income tax	(47 000)	21 508	(58 300)	(42 201)	(27 501)	(47 649)
Net profit	62 243	185 672	147 078	87 429	73 959	11 782

Sources: *Australian Financial Review Shareholder*, 13th edn, Ian Huntley Pty Ltd, 1998, p. 216; and John Fairfax Holdings Ltd Annual Report, 1998, p. 41.

cost cutting, and the drive for greater syndication has been at the expense of diversity and opportunities for newcomers. They have seen eleven chief executives or chairmen in the eleven years since Warwick Fairfax's disastrous buy-out attempt in 1987. The company's financial performance during recent years is summarised in Table 3.6.

Like most established media corporations, Fairfax is also searching for a commercially credible Internet strategy. Understandably, the company does not want to initiate any strategy that would undermine its 'rivers of gold' classified advertising base. Speaking at the group's annual meeting in November 1998, Chairman Brian Powers said that Fairfax's on-line investments, which included its Internet home pages and classified sites, lost about $10 million in 1997/98 and could lose a similar amount in 1999. He added that the company would continue to invest in the Internet, 'even if it never made a penny', because the initiative 'enhanced our company in an increasingly competitive and converging market place'.[9] Later, Fairfax's Managing Director, Fred Hilmer (the government's architect of competition policy), said that floating the on-line division, which doubled its revenues in 1999 but still remained unprofitable, was not a priority. In contrast to PBL, Hilmer argued: '[W]e are focusing on building a viable on-line business and we think this will determine the financial structure of the business. Not the financial structure determining the business.'[10] The Hilmer strategy is to keep all future Fairfax Internet options open. 'Each of our bets is a long shot,' said Hilmer, 'and I don't know which one will come off—whether it will be transaction revenue that works, or subscription revenue, or banner advertising or classifieds—or some combination of them all.'[11]

Ultimately it may be what Fairfax has been best at in the past—substantive content—that could see it emerge as one of the strongest commercial Internet forces in Australia in the future.

Kerry Stokes at the Seven Network

Seven's television interests have been reconfigured in recent years around the business strategy of Kerry Stokes, who has extended his media league from ownership of the *Canberra Times* and the Golden West Television network to the chairmanship of the Seven Network. During this period he has increased his personal shareholding from 20 per cent to 33 per cent in 1999. Investors and the financial press have been scathing in their overall assessment of the financial performance of Seven under Stokes. Luke Collins of the *Australian Financial Review* contrasts the previous financial performance of Seven in the last full year under then Chairman, Ivan Deveson, and former Managing Director, Bob Campbell, when Seven posted a net profit of $90.2 million on sales of $555 million to the year ended 30 June 1994, with the financial performance under Stokes in the 1997/98 financial year of a profit of a mere $20.4 million on revenues of $800.2 million.[12] There had been continual speculation that Seven's eight institutional investors—Colonial, Portfolio Partners, Permanent Trustee, Deutsche Australia, Lend Lease, State Super, Country and National Australia Trustees—had put the other one-third owner, the Seven Network Chairman, on notice. In July 1999 Stokes took over as Executive Chairman and announced that he planned immediate cost reductions of A$30 million a year and planned to rebuild the core television business.

Seven spent US$250 million on a major shareholding of the ailing MGM film studio, in partnership with an American industrialist, although after incurring a loss of about A$100 million in MGM, Seven dumped this investment. Seven also purchased the risky Australian Television commercial satellite television venture for the Asia Pacific region from the Australian Broadcasting Corporation but it quit this operation in June 1999. Seven was also a 30 per cent stakeholder in Sports Vision, an Optus pay television initiative, but having devoured an estimated $4.5 million in equity and a further $17 million in loans, the Seven Network board stopped supporting the company in 1997. But the chief criticism of its financial management is that it pays too much for its television programs for too little return.

However, not all of its financial news has been bad. It acquired a small equity in Optus Communications, with rights to 50 million shares at $1.25 per share, and an entitlement to acquire an additional 50 million shares when the company was listed on the Australian Stock Exchange. On debut day for the renamed Cable and Wireless Optus, 17 November 1998, Kerry Stokes' Seven Network Ltd was sitting on a paper profit of $87.36 million for its investment. In March 1999 Seven announced that it had sold these shares at $3.65 each for a gross profit of $130 million. Most of the criticism of the Seven Network centres on how management runs its core television business. Its financial performance during recent years is summarised in Table 3.7.

Table 3.7 Seven Network Ltd: Balance sheets, 1993–98 ($m)

	1992/93	1993/94	1994/95	1995/96	1996/97	1997/98
Sales revenue	564.4	555.0	582.8	671.2	756.9	800.2
Income tax	–	–	–	(3.3)	(51.6)	(38.7)
Net profit	47.5	90.2	41.3	115.1	88.9	20.4

Source: The Seven Network Annual Reports, 1997 and 1998.

Channel 7 has long been regarded as Australia's leading sports television network, the 'footy and Olympics channel'. Overall, when Channel 7 won a ratings week in Melbourne in September 1998, it was the first time it had led in 80 weeks; since early in 1991, its arch rival, Channel 9, had won 270 of the 308 official Melbourne audience surveys. Seven has been unable to command strong prime time audiences for drama, although notable exceptions are its flagship Australian production, 'Blue Heelers', and its best import, 'Ally McBeal'. It has a high dependence on Australian Rules football for its programming. Building its own form of vertical integration, the Seven Network committed $99.5 million in September 1997 to the development of the Docklands Stadium Consortium for the Australian Football League (AFL), gaining Seven the rights (over PBL) to the purchase of all premium seating in the stadium, and the naming, signage, advertising and ticketing rights to all events in the stadium for 25 years. A common estimate is that it will take Seven ten years to recoup its stadium outlay. As well, Seven has secured free-to-air television rights to AFL matches from the 2002 season for $10 million and has committed an additional $10 million in AFL sponsorship. The convergence of media and sports has become big business in Australia.

Seven's diversification into new media appeared to be emerging from a joint venture with the computer chip giant Intel. Attempting the innovation of watching television on a personal computer and receiving web pages over the airwaves by connecting their Intel card to their television antennae, Seven planned to provide a 'one-stop shop' with its television content, AFL web site and other network Internet services. Inspired by the success of the AFL web site, which sometimes records a million 'hits' a day, Graham McVean of Seven predicted that this new integration would mean that 'web surfing will be a thing of the past'.[13] Overall, though, there does not appear to be a clear sense of a convincing new media strategy in Seven's planning, just general reference to 'PC on TV' and the 'TV on PC' in the future. Its Chief Executive, Julian Mountier, suggested just before his resignation in July 1999 that the Seven Network planned to launch an interactive television service and was also considering developing a share-trading site.[14]

The common financial marketplace assessment of Seven is that it lacks an integrated programming strategy, that it is commercially vulnerable because of its overdependence on sport, especially AFL football, and its costs are not tightly controlled. Seven, though, seems prepared to continue to chance its programming arm more than some of the other networks while consolidating its sports and entertainment complex base. The network has yet to win the real imprimatur of the Australian investment community. It just may be that Kerry Stokes, new boy on the block of the big-time commercial network TV and invited ABC Boyer lecturer, is just a little more complex than others in his motivation for being Chairman of the Seven Network.

Canadian capitalism: The Ten Network

Ten Network Holdings Ltd is a case of remarkable financial revival in recent years. Late in 1990, in the aftermath of the debacles of the crazy television takeover frenzy, Westpac bank was forced to take over the Lowy family debt associated with their $700 million purchase of Channel 10 which was losing an estimated $2 million a week. Ten was considered to be the basket case of the Australian commercial media. In 1992 a group of investors led by CanWest, a Canadian television broadcasting company, acquired the Ten business from Westpac. In 1995, with the purchase of affiliated stations in Adelaide and Perth, Ten secured all five major Australian television markets, reaching 65 per cent of Australia's population. CanWest then held 57 per cent of Ten through debentures, well above the 20 per cent foreign ownership limit, but the Australian Broadcasting Authority (ABA) approved this on the grounds that it held only 15 per cent of the voting shares. In November 1996 and January 1997 CanWest acquired an additional $200 million of Ten shares from other investors, lifting its economic stake in Ten from 57 per cent to 76 per cent, although CanWest claimed that its voting interest remained at 15 per cent. However, in April 1997, the ABA ruled that CanWest was in a position to control the holding companies, giving it a voting stake of 52.5 per cent and that it was thereby in breach of the foreign ownership rules. Treasurer Costello ordered the five companies financed by CanWest which lifted their financial interest in Ten from 57 per cent to 76 per cent to divest those shares within six months, regardless of price.

In March 1998 Ten Network floated on the Australian stock market and investors scrambled to get their hands on the $356 million worth of shares on offer. Ten was valued at $2 billion at the float price of $2.15 a share, and CanWest was reported to have pocketed $120 million on its shares for the 18 per cent it sold into the float.[15] A neat profit for a Canadian company on the back of Australian public assets—television licences—which still left CanWest with its 57 per cent interest in Ten. Izzy Asper, Canadian CEO of CanWest, had 'done a Murdoch in Australia'.

The Ten Network management formula has been clever financially under CanWest's influence. Their formula has been to tailor programming to their core demographic audience, especially the 16–39-year-olds, with related imported programs such as 'Melrose Place', 'Beverly Hills 90210' and the 'X-Files', as well as imported comedy programs such as 'Seinfeld', 'The Simpsons' and 'Mad About You'. Ten has stayed away from the high-investment programming areas which directly compete with the Nine and Seven networks, has opted for comparatively cheaper local programming, such as the light but accessible current affairs chat show, 'The Panel', and stole the zany 'Good News Week' from the ABC, stayed out of blockbuster sporting investments like cricket or the AFL Docklands deal, delivered discounts to its advertisers, and has been ruthless in cutting its general overheads. The Ten commercial model has had a profound effect on the corporate financial strategy of the Nine Network—and the future of Australian commercial television. The cost reductions clearly impact on the balance sheets for Ten Holdings Ltd (see Table 3.8).

Ten Network seems likely to stick to its television knitting. Its web site is a token gesture, but as CEO McAlpine has explained, 'If Bill Gates isn't sure of the outcome [of his on-line joint venture with Nine] then I don't think anyone should be. There's plenty of time to walk through this one.'[16]

Three conclusions can be drawn from this analysis of Australia's commercial media club. First, this group of major media stakeholders is one of the most ruthlessly commercial Australian industry sectors. Second, each of these organisations has experienced extremely strong revenue growth during the past half decade, at a time of low inflation, although their profitability performances vary widely. Third, the source of their growth during this period remains basically centred around the

Table 3.8 Ten's Consolidated Holdings (excluding Adelaide and Perth prior to their acquisition): Balance sheets, 1994–98 ($m)

	1994	1995	1996	1997	1998
Gross revenue	377	398	478	548	552
Income tax	15	14	24	35	35
Net earnings	55	40	56	66	72

Sources: Ten Holdings Ltd Prospectus, 4 March 1998; and 1998 Annual Report.

'old media', notably classified advertisements in newspapers and advertising revenue from prime time commercial television. None of these groups has any clear sense of a future development strategy for new media.

Adding a wheel: Pay television in Australia

Subscription television, more commonly referred to as pay TV, has been compared to a car with five wheels—the car may be new, but what does the extra wheel really offer? An initial competitive battle in Australian telecommunications was waged around the introduction of pay TV, a sorry political and corporate saga of huge losses. This service was inevitably going to involve corporate alliances because of the convergence of broadcasting companies, who have the programming, with telecommunications carriers, who have the distribution systems to deliver to customers. With conventional broadcast television, anyone with a television set can access the service, but pay TV is discriminatory as the user pays, requiring the company to purchase and market its pay TV services, distribute its programs to select customers via a network, and bill them directly. Although pay TV has been well established in North America for decades, and much of Europe for more than a decade, Australian governments long procrastinated over its introduction. For years, despite significant advocacy on its behalf and detailed policy inquiries, notably a five-volume report in August 1982 when the Australian Broadcasting Tribunal recommended its introduction 'as soon as practicable', pay TV in Australia has been dogged by difficulties, delays and strong opposition from established vested commercial television interests.

Pay television stumbled badly after its belated introduction into Australia. Australis Media, using microwave multi-point delivery systems, was the first company to launch pay TV services in Australia in 1995. After heavy trading losses ($251.7 million for 1996 and another $297.5 million loss for 1997[17]) and two merger attempts disallowed by the Australian Competition and Consumer Commission (ACCC), Australis Media Ltd was finally placed in receivership in May 1998. ACCC Chairman Allan Fels refuted that any blame for the failure of Australis lay with his organisation, saying: 'Australis paid $184 million to get its pay TV licence, it bid extremely high prices for Hollywood movies by international standards and it supplied expensive pay TV services for well below cost.'[18]

The impact of the introduction of pay TV was always seen as having wider industry effects than merely the introduction of a new competitor into Australia's television market. The decision by both Optus and Telstra to enter the subscription market in Australia was to some extent a 'loss leader' telecommunications strategy. Telstra was concerned that newcomer Optus might be able to build a significant customer base through its own pay TV network which could be used to build up its market share of the local telephony markets, and later for emerging broadband services. Optus, of course, wanted to deliver local telephony and did not want to remain dependent on Telstra's network to access its own customers. Optus, however, experienced considerable technical problems in offering its own local telephony service over its pay TV network, much to its chagrin.

Initially, the two carriers intended to invest $3–3.5 billion each to roll hybrid fibre-coaxial cable past most metropolitan homes by 1998, but these plans stalled in 1997 due to staggering losses. Telstra's pay TV collective losses were estimated at $800 million by August 1997, and the full-year Optus Vision loss was reported at $387.9 million, a significant contributor to parent company Optus Communications' loss of $411.4 million for the year ended 1996.[19] Telstra's cable passed 2.1 million homes (the original plan was four million) and Optus passed 2.2 million households (three million originally planned).[20] Not even the free-wheeling capitalist United States had invested in dual cable infrastructure, but Australian public policy belief in competition enabled Telstra and Optus to construct separate broadband networks to service the same homes—the duplication level was 85 per cent![21] Key rivals Foxtel and Optus Vision agreed on a satellite sharing deal in December 1998, but they did not come to a programming arrangement which would have allowed subscribers to receive programming from more than one pay TV operator on the one service, as some had advocated. The take-up of these services has improved and, as Table 3.9 shows, had reached respectable numbers by 1998.

Table 3.9 Australian major subscription television operators and take-up, 1998

	Cable	Satellite	Microwave
Foxtel	360 000	40 000	–
Optus Vision	200 000	–	–
Austar	5 000	200 000	80 000
TARBS	–	–	70 000

Source: *Australian Business*, 24 December 1998, p. 21.
Note: TARBS is the ethnic broadcaster Television and Radio Broadcasting Services Pty Ltd which owns microwave set-top boxes in those homes which subscribed to the now defunct operator, Galaxy.

In 1998, stockbrokers J.B. Were neatly summarised the Australian pay television outcomes to date:

- Multi-billion dollar operating losses.
- Duplication of capital infrastructure designed to facilitate improved competitive positioning in the broader telecommunications industry.
- Unsustainable levels of pay TV customer churns, due to an inability of operators to meet customer requirements created by their own aggressive marketing campaigns.[22]

There has been a great deal of speculation about major rationalisation in the Australian pay TV industry. Although the rate of subscriber take-up has actually been favourable by American and European historical comparison, Australia's pay TV industry has lost billions of dollars—through the collapse of Australis Media, and huge write-downs by Foxtel and Optus Vision—and no operator has made a profit in its Australian history. To resolve the difficulties, an alleged agreement of June 1997 is often quoted between some of the principal media stakeholders, including that News Corporation and Publishing and Broadcasting Ltd would 'equalise' their interests in pay TV.[23] Subsequently, in September 1998, PBL spent $150–160 million to become a 25 per cent owner of Foxtel. This gave both Murdoch and Packer 25 per cent of Australia's most successful pay TV company, Foxtel, together with Telstra as owner of the other 50 per cent.

Perhaps because Australia is a comparatively small pay TV market, an industry rationalisation makes sense. A pay TV duopoly would reduce the losses, deliver about 600 000 subscribers to the operator, consolidate the diffuse programs they now offer, and probably reduce the price they currently separately pay for Hollywood movies and sports programs. The fees paid by consumers could be monitored. Given the losses to date, it is hardly surprising that the initial plan that pay TV drama channels should spend 10 per cent of their annual program expenditure on new Australian drama has not been met. With a more secure financial base, this quota could be increased to promote local drama production.

However, the ever vigilant Allan Fels, Chairman of the Australian Competition and Consumer Commission, has warned that 'the ACCC would be concerned if the alliance of the interests of PBL, News, and Telstra were used in future to lessen competition in Pay TV and related broadcasting and telecommunications markets'.[24] Later, in a move designed to break up a possible Telstra and Cable and Wireless Optus duopoly in pay TV, the ACCC issued a draft decision to 'declare' that owners of pay TV facilities must allow access by third parties. This decision was made after ethnic pay TV operator, Television and Radio Broadcasting Services Pty Ltd, requested access under the provisions of the *Trade Practices Act*.

The losses are likely to go on in this relatively small pay TV market. Yet, as we are so often told, competition brings the best out of both people and companies.

Identity revisited: Public broadcasting

Our media and communications system has long been built around a mixed economy of public and private sectors, but with significant public institutions and also government intervention in certain practices, defended in legislation on the grounds of being 'in the public interest'. Yet what does this term mean? Dennis McQuail suggests that the word 'public' indicates 'what is open rather than closed, what is freely available rather than private in terms of access and ownership, what is collective and held in common rather than what is individual and personal'.[25] Critics argue that the use of 'public interest' actually comes down to someone's version of what is in the general good.

The rationale for public broadcasting is situated in the enlightened notion of a public sphere where vibrant political life and social interaction unfold for the whole community. Broadcasting is one of the great outcomes of democracy, and public broadcasting addresses and promotes the rich diversity of a good democracy.

Our favourite aunt: The Australian Broadcasting Corporation

Public broadcasting had its origins in Australia in the 1920s when there was a perceived need for broadcasting services to be established which had declared institutional independence and freedom from political interference. The Australian Broadcasting Commission (ABC) came into being under the *Australian Broadcasting Commission Act* of 17 May 1932. On 1 July 1932 the then Prime Minister, Joseph Lyons, said in the ABC's first broadcast that the commissioners' task 'would be to provide information and entertainment, culture and gaiety, and to serve all sections and to satisfy the diversified tastes of the public'.[26] The then Chairman of the British Broadcasting Corporation (BBC), Charles Lloyd Jones, observed that the ABC was a fine set of initials for a body setting out to enlighten the nation, connoting the beginning of knowledge and 'forming an appropriate co-relative to the abbreviation BBC'.[27] The ABC's historical location is firmly within the tradition of its British counterpart, and the BBC's declared pride in a commitment to truth and the pursuit of excellence in broadcasting was inherited by the Australian imperial artefact. There has long been a sense within the ABC of its own importance as a national institution, as a kind of central moral authority in Australian society.

The foundations of public broadcasting are situated in how a democracy caters for its cultural and political diversity. Commercial broadcasters have to deliver the largest possible audiences to their advertisers, as measured each quarter of an hour by the ratings. However, the role of the public broadcaster is that they must relate to audiences as a public, rather than a market. While talented public broadcasters will always produce and present programs that will capture significant audiences in size, and capture audience heartlands with ideas and imagination, they must also offer specialised progams which address the diversity of interests and tastes of their audiences. Hence their charter is to be comprehensive in programming, unlike commercial broadcasters, yet critics of public broadcasting have long labelled their programming as being 'elite' or 'minority' or 'alternative' or patronising 'niche' offerings.

The view of the inherent worth of public broadcasting and the importance of the struggle to maintain it institutionally within a vibrant public sphere is far less widely accepted today. Commercial broadcasters tend to see public broadcasters as a self-serving, self-interested group who construct programs which tend to reinforce their own view of what the world is like, or what it should be like. For postmodernist academic Ien Ang, the entire history of the BBC should be seen in terms of audience resistance to the paternalism of the BBC's cultural mission.[28] Ang argues that the Reithien model of public broadcasting (John Reith was the BBC's most illustrious Managing Director who started his career in 1922) articulated the values, standards and beliefs of the British upper middle-class in a public broadcasting system which was a form of enlightened cultural dictatorship. Ang argued that what Reith strived for 'was the creation of a common national culture: the BBC's self-conception was that of a "national church" (Reith's own words) to whose authority all citizens should be subjected'.[29] Commercial broadcasting was introduced much later in the United Kingdom than in most Western democracies, and for Ang there is now, as Goode explains, 'the pragmatic acceptance of a market rationality which then sits uncomfortably alongside the residues of traditional elitism and state control'.[30] Alternative perceptions of reality may not necessarily depend upon the existence of big, cumbersome bureaucratically organised programming factories whose energies too often focus on protecting their institutional patch in society.

So, what assessment may be offered of the role of the ABC today? At one level, the ABC is a remarkable hybrid organisation which provides Australia-wide radio networks, including Radio National, ABC Classic FM, the FM youth network, Triple J, Radio Australia, and the Parliamentary and News Network, as well as metropolitan and rural radio stations which reflect regional diversity. According to a survey conducted by Newspoll, ABC radio held its position in a highly competitive market with a 21 per cent audience share in 1998, up 4 per cent since 1991.[31]

ABC TV, Australia's only national television service, is carried on over 600 transmitters. Its programming includes its flagship trusted daily news programs, together with quality exploratory current affairs, such as 'Four Corners', 'The 7.30 Report', and 'Lateline', as well as innovative programs in recent years across different genres, such as the drama of 'Phoenix', 'Janus' and 'Wildside', the widely appealing 'SeaChange', the brilliant media satire of 'Frontline', the comedy of Roy and H.G. ('Club Buggery') and 'Good News Week' (subsequently poached by Channel 10), science frontiers on 'Quantum', and 'Playschool' and 'Bananas in Pyjamas' for children. According to the Newspoll survey above, the ABC turned in its strongest performance in 1998 in terms of television audience ratings in both regional and metropolitan markets by earning an average 15 per cent of prime time audiences throughout Australia.

ABC Enterprises offers the ABC Shops and ABC Centres, for books and magazines, classical music, audio cassettes and video services. And ABC Online provides the most accessible text service to its programs via the Internet of any Australian broadcaster. According to the Newspoll survey above, ABC Online was the most popular Australian media Internet site in 1998 with an average of about seven million accesses per month—up 50 per cent in a year. The ABC has clearly led Australian broadcasters in showing what can be achieved by the migration of broadcasting services to the on-line environment.

Wherever public broadcasting is organised as some form of public service institution heavily reliant for its revenue on the government of the day, the organisation is inevitably dependent upon political circumstance. The ABC, a public institution with a budget of over half a billion dollars annually, must, of course, be rigorously accountable to the Parliament and to the people for their annual expenditure of substantial public funds. Yet public broadcasters need a great deal of editorial freedom and independence in order to do properly what they were established to do. No matter what formula operates to allocate public funds to public broadcasters, be it direct government appropriation as for the ABC, or licence fees as for the BBC, the organisation is financially in the hands of the government of the day, and how government perceives their performance. Finding the proper relationship between financial accountability to the government of the day, together with the capacity to maintain fierce programming independence from the same master, has long been a key unresolved dilemma for public broadcasters. The notion of a truly independent or autonomous public service broadcasting institution is illusory.

In the current re-evaluation of the role of the state and its benefits, public broadcasting has found itself no longer readily accepted as serving comparable functions such as state education or health care. In a world with competing claims for diminishing public resources, Britain's BBC, the Canadian Broadcasting Corporation (CBC), Japan's Nippon Hosa Kyokai (NHK), Australia's ABC and the Special Broadcasting Service (SBS) have all found themselves faced with severe financial cutbacks from their political masters.

Most recently, the ABC has found that government cutbacks have forced substantial programming rationalisation on the organisation, with a reduction in its government funding by $11 million in 1996/97 and by a further $55 million in 1997/98.[32] This contributed to the closure of some of its international short-wave radio broadcasting service, Radio Australia, and to the incorporation of both of its Melbourne and Adelaide symphony orchestras from 1 July 1997 with the remaining orchestras, who had collectively provided 700 performances a year to over a million people.

Table 3.10 shows the income plateau of revenue made available to the ABC from government appropriation during the past five years.

While financial comparisons between commercial and public broadcasters cannot be easily made, the general financial trends during the past five years show that the ABC has fared badly in recent years (see Table 3.11).

Several conclusions can be drawn from the above data. Clearly the ABC's financial base is non-growth at a time when two of its major rivals in television, the Seven and Ten networks, have grown by more than a third during the past five years, and PBL, which includes the Nine Network, doubled its revenues through the period 1993–97. The Ten Network, widely touted as 'the mean and lean' organisation, had a revenue base of $553 million in 1997 for television, whereas the whole of the ABC, with its national television, radio and orchestral responsibilities, had a revenue base of $588 million.

Table 3.10 The Australian Broadcasting Corporation: Five-year financial analysis ($m)

	1993	1994	1995	1996	1997
Cost of service	671	688	699	720	707
Operating revenue	90	101	123	126	132
Net cost of service	581	587	576	593	575
Revenue from government	576	602	588	598	588

Source: ABC 1997 Annual Report, p. 6.

Table 3.11 Australian major media: comparative revenue base, 1993–97 (in $m, except percentages)

	1993	1997	% growth
ABC	576	588	0.02
Fairfax	771	1027	33.2
PBL	562	1201	118.7
Seven	564	760	34.7
Ten	377	553	46.7

Source: See data from Annual Reports in Tables 3.5, 3.6, 3.7, 3.8, 3.10.

Morever, the increasingly globalised nature of broadcasting, together with the explosion of new delivery systems and increased channel capacity, has radically altered the contemporary broadcasting environment. How does a publicly funded broadcasting institution procure the substantial capital injections necessary for major new investments in new technology? And if it chooses to ignore the new delivery environments, might the ABC find its single channel television offering eventually swamped in any city which may offer its consumers up to 500 channels in the future? There are many strategic questions concerning whether a public broadcaster should be in the new technological game of pay television, global networking and the Internet, and how they can possibly fund such developments.

Public broadcasters have been called upon to define themselves in the new monetarist context and justify their very existence. What ought to be the role of public broadcasting in contemporary society? The answer depends upon what communities expect of public broadcasting today, whether they see its goals as worth continuing support, and how the political system decides about its future. Canadian academic Marc Raboy writes, '[I]n the new broadcasting environment, the issue of public service broadcasting can be reduced to this: What social and cultural goals attributed to broadcasting require a specially mandated, non-commercially driven organisation, publicly owned, publicly funded to the extent necessary, and publicly accountable?'[33] One attempt at addressing the role of public broadcasting in contemporary society was made by the UK's Broadcasting Research Unit (BRU), itself now defunct. The BRU argued that public broadcasting should still continue to strive for comprehensiveness rather than merely state which of its functions deserve to be retained in the new environment. They summarised the following main principles of public broadcasting:

- universal accessibility (geographic);
- universal appeal (general tastes and interests);
- particular attention to minorities;
- contribution to sense of national identity and community;
- distance from vested interests;
- direct funding and universality of payment;
- competition in programming rather than for numbers; and
- guidelines that liberate rather than restrict program makers.[34]

These have long been widely accepted attributes of the public broadcasting system, although the policy winds of change are now calling for the abandonment of comprehensiveness as the centrepiece of its *raison d'être* in a shift towards narrower and less expensive goals.

Brian Johns, ABC Managing Director, has argued that the ABC has undergone many metamorphoses, but has survived to remain among the most respected institutions in Australia. He addressed the issue of the source of its vitality in summarising his view of where the ABC sits in Australian society today:

- a mandate for creative exploration in an environment where commercial interests do not determine schedules;
- a deep rooted presence across the country, and the competitive edge which that diversity of experience brings to its programming;
- an enduring role in educating Australians and increasing the awareness of the community as a whole;
- a proud Australian culture and a vigorous creative community which support and are supported by the ABC;
- a comprehensive editorial brief—allowing the ABC to range over the gamut of issues and program styles of interest and relevance to our audiences;
- the watchdog Australian public—quick to recognise mediocrity and superficiality, but equally quick to defend the ABC from political or commercial interference.[35]

It is unfortunate that so few comparable attempts have been made within private sector broadcasting organisations to articulate what they stand for.

The re-evaluation of the role of public broadcasting in contemporary society has also been subjected to formal inquiry. At times during the 1980s and 1990s, when private broadcasting was temporarily in a shambles because of media takeovers, conservative governments somewhat ironically established major national inquiries into the public sector, the ABC. In May 1981, Alex Dix, Managing Director of Reckitt and Colman, in association with a four-person committee, issued their five-volume report, *The ABC in Review: National Broadcasting in the 1980s*. Rather than offering any criticism of the Fraser government's financial cutbacks to the ABC, Dix was in fact more critical of the ABC's response to them. Although Dix acknowledged the many talented and dedicated people who worked for the organisation, he argued that the ABC had faltered in the past decade and 'stood on its dignity and independence when pressing problems cried out for attention'. The Dix report certainly did not portray the ABC as an innovative broadcasting pacesetter, but rather as an organisation which had become 'slow moving, overgrown, complacent and uncertain of the direction in which it was heading'.[36]

For Dix, the corporate structure of the organisation, and its associated management capability, was at the heart of the problem and 'the objective of many of our recommendations is to structure the ABC more like a commercial organisation'. Dix's report was seminal in that it led to the emergence of a different kind of institutional ABC. Although his recommendations sat on the shelf for some time because the Fraser government was caught up in the 1983 election, which it lost to Labor, the new Hawke administration moved quickly in 1983 to implement the spirit of the changes advocated by Dix. The Australian Broadcasting Commission became the Australian Broadcasting *Corporation*, and in an important symbolic gesture Labor chose Ken Myer, elder son of Sir Sidney Myer, founder of the retail empire, as its inaugural corporate Chairperson. The new management structure was that of a board of six to eight non-executive board directors, who appointed a managing director to head an organisation released from the control of the Public Service Board, though not from restrictive clauses on investment and borrowing. ABC management had been privatised in spirit though not in substance.

Whereas the Dix inquiry had set in place the corporatisation of the ABC, the formal inquiry conducted in 1996 by Bob Mansfield, formerly the inaugural CEO of Optus and later Fairfax, recommended its extension, although neither Dix nor Mansfield recommended the introduction of commercial advertising. In a more personalised though most professional investigation, Mansfield, like Dix, attacked the ethos of the organisation that ABC management had created. According to Mansfield, '[T]he ABC does not have the feel of a modern, dynamic organisation. Its lack of willingness to embrace new ideas is inconsistent with its role as a creative organisation and will ultimately retard its creative capacity if attitudes do not change.'[37] Mansfield's report was not as far-reaching as Dix's, but its most provocative recommendation was 'that the ABC move to outsource the majority of its non-news and current affairs television production over the next three years'.[38] Here he drew upon his discussions with the BBC, arguing that substantial savings could be realised from a rationalisation of properties owned by the ABC and by examining alleged production overcapacity. On release of the report, ABC Managing Director Brian Johns was most critical of Mansfield on this point, arguing that the international experience in leading contemporary broadcasting organisations was actually shifting in the opposite direction, with major investments in new in-house production capacity. Lisa Fowkes, union secretary, warned that this recommendation 'could cost up to one-fifth of all ABC jobs from the present work force of 5000' and suggested that the Mansfield report 'had delivered what the government wanted from day one—a smaller, compromised and weaker ABC with its creative heart removed'.[39]

Behind this notion of a shift to major programming outsourcing is the seeds of a sea change in terms of the identity of the ABC—a highly politicised issue. Phillip Adams once said of the ABC that if it was an art gallery the sign out the front would have to say 'only staff portraits hang here!'. There is a fundamental question here as to whether the programming modus operandi of the ABC could be opened up to wider participation, despite the perceived threats to ABC staff and their sense of professionalism, to foster more ideologically diverse, alternative and antagonistic discourses in our media system. This, of course, is a discussion that at least we can have about the future policy of the public's ABC, but one that is hardly possible to have about a commercial media organisation.

It is remarkable how there is so much discussion about the alleged bias of the ABC, especially during election campaigns. Bias, of course, is inevitable in any broadcasting. How could anyone produce a value-free radio or television program, or newspaper, within either the public or private sector? The gathering, editing and publishing of news involves decisions by people who inevitably bring their own backgrounds, values and prejudices to bear on deciding what to select, emphasise and colour as news and current affairs. We cannot expect any broadcaster to produce 'unbiased' programs, but we can expect them to apply the historian's sense of *fairmindedness* across the range of their programs.

During the 1998 election campaign the ABC was again subject to the common outcry of 'left-wing bias' for its alleged pro-ALP stance. Newly appointed ABC board member and Liberal Party power broker Michael Kroger wrote an opinion piece, 'Why the coalition is fed up with ABC bias',[40] where he argued that 'whatever procedures there are to insure impartiality and a lack of bias do not seem to be working'. Apparently, the Coalition made seventeen formal complaints to the ABC alleging bias and unfair treatment during the campaign. Some politicians who accuse the ABC of being a spent force rather oddly also seem to believe that the ABC is a highly influential force in shaping public opinion, especially at election time! Meanwhile, ABC staff alleged bias of a different kind when ABC Chairman Donald McDonald eulogised the Prime Minister, John Howard, at a Liberal Party fundraiser. This cut and thrust is all part of a healthy media environment. The time to really worry about the value of the ABC is when there is no allegation and counter-allegation of alleged bias in vigorous discourse.

A further pressure on the ABC is how it responds to this new globalised multi-channel digital world of technological plenty. It is financially locked out of direct advertising revenue, as it should be to retain important qualities of its identity and independence, yet new ventures such as pay television, digital television and international broadcasting demand substantial capital investment. The ABC has to run cap in hand to the government of the day for any major strategic capital investment for the future. Unfortunately though, where ABC management has acquired such special purpose development funding, the outcomes have not been productive to date, notably with their new international television service and their initiative in subscription television programming.

In 1992, federal Cabinet allocated $5.4 million to ABC TV to operate a satellite service, Australian Television International (ATVI), primarily to South-East Asia. Although President Ramos of the Philippines once said that ATVI provided the best news service in Asia, it did not reach the heights of its counterpart, BBC Asia; the project floundered financially and was sold to the Seven Network. According to Ken Lenthall, ATVI 'grossly overspent its allocation, accepted sponsorship, and continued to lose money, while not attracting a substantial Asian audience'.[41] Such a project, however, was always going to be difficult (Murdoch's Star TV service in Asia has never made a profit), and the ATVI service needed much more time to establish itself. It is difficult for a public broadcaster to go back to its financial master asking for recapitalisation of ventures that are inevitably long term.

Shortly after the ABC's Asia Pacific venture was launched, management also embarked upon a subscription TV venture named Australian Information Media (AIM). Clearly, the ABC was never going to be a systems operator in pay TV, like Foxtel or Optus Vision, so it conceived its role as offering a specialist news channel building on its programming strengths with the intention that this would run on the pay TV networks. With a government grant of $12.5 million, new state-of-the-art studios and a staff of 100 working on the venture, the initial signs were promising. AIM was important in positioning the ABC in the emerging new media structures. However, all three major pay TV systems operators—Foxtel, Optus Vision and Galaxy TV—rejected the fledgling service on the grounds that it was too expensive. Hence, they had locked the ABC out of the new pay TV services. The full tale of the terms of this rejection has never been properly told, least of all to say that no love was ever lost between the commercial operators and the ABC in major strategic innovation.

So, *whither* the ABC? We ought to consider in our assessment of the ABC its contribution to the Australian media sector as a whole, where it remains as a major training ground for commercial media personnel. It has long had, and continues to have, the top stable of journalists, international correspondents and program presenters in the nation, as good as the best anywhere in the world. This line-up includes (in alphabetical order), Phillip Adams, Monica Attard, Jennifer Byrne, Quentin Dempster, Jon Faine, Jonathan Holmes, Terry Laidler, Terry Lane, Chris Masters, Kerry O'Brien, Greg Wilesmith and Robyn Williams. What a training ground the ABC has provided for commercial networks over the years, including the now richer Paul Barry, Richard Carleton, Ian Leslie, Ray Martin, Michael Willesee and, more recently, Maxine McKew and Ellen Fanning. The ABC, one of Australia's most loved and respected institutions, finds itself constantly under political siege and faces great pressures, financially and managerially. Yet no matter what the threat, Australians will come out of the woodwork from all walks of life to protect the ABC. Sir Rupert Hamer and June Factor, on behalf of Friends of the ABC, said in 1997:

> Friends of the ABC are once again roused to defend our remarkable 64-year-old national broadcasting enterprise of quality and integrity. The ABC belongs to the Australian people, who willingly fund it. Governments are the ABC's guardian, not its owner.[42]

Perhaps we ought to revisit the rationale for its fundamental existence rather than try to create for it a new false identity. Is there a convincing argument that the ABC is an anachronism that needs to be put down?

There is important evidence that it has long satisfied Australian audiences, and continues to do so. In the early 1980s Alex Dix reported that 'we know Australians want a national broadcasting service. Those giving us their views stressed the importance of the ABC in their lives, and, by implication, in the shaping of the kind of society in which they want to live.'[43] Similarly, Bob Mansfield (formerly Optus's inaugural Managing Director), reported in the late 1990s that 'most private companies would envy the passion and loyalty which characterises the relationship of the ABC with its audience—it is clearly a special relationship'.[44] Yet despite the remarkable rapport and respect that the ABC has with millions of its devoted followers, there is constant sniping from its commercial competitors, and disgruntled staff, about almost every aspect of its operations. Paradoxically, the ABC is Australia's most loved and most maligned national organisation.

Ultimately, we need to evaluate the contribution of the ABC not in terms of a management structure or a funding formula, or the size of this or that audience, but in terms of what it contributes to the way audiences understand their world and enjoy life. Friends of the ABC have suggested that it is within the public sphere that our sense of unity in diversity can be maintained in Australian society. The richest country in the world, the United States, is so much the poorer for not having a vibrant mainstream public broadcasting equivalent to the Australian Broadcasting Corporation.

SBS et al.: The Special Broadcasting Services

Australia is one of the few countries in the world with two public service broadcasting organisations, the ABC and the Special Broadcasting Service (SBS), separate organisations each with distinct charters. In 1974 the Whitlam government approved two government-funded radio stations in Sydney and Melbourne, 2EA and 3EA respectively. In September 1976 Prime Minister Malcolm Fraser, a long-time champion of multiculturalism, invited the ABC to establish a permanent ethnic broadcasting service, but withdrew the offer in June 1977 and created instead another statutory authority, SBS, as an election promise which came into being in January 1978. Australia's first multicultural television channel began its operation in Sydney and Melbourne on United Nations Day, 24 October 1980.

SBS has always lived an uneasy existence. It has attempted to establish a national service like the ABC, but with more modest resources; its radio signal with 68 languages is broadcast to all capital cities, and SBS television nominally covers 80 per cent of the population. This is achieved on an annual budget of $95.2 million (in 1997), about one-sixth of that of the ABC, of which government appropriation amounted to $83.6 million and advertising and sponsorship raised $11.9 million. It employs around 900 full-time and part-time people.[45] Whereas commercials are prohibited on the ABC, advertising and sponsorship announcements have been allowed on SBS Radio and SBS TV since 1991 for up to five minutes an hour.

SBS boasts an innovative television diet, including the most internationally focused nightly evening news service, two probing current affairs programs, 'Dateline' and 'Insight', a nightly sports program, 'Toyota World Sports', lifestyle programs, 'Second Opinion' and 'The Food Lover's Guide to Australia', and 'The Movie Show'. SBS TV broadcast almost 2000 hours of locally produced programs in the 1995/96 financial year on shoestring budgets, equivalent to 31 per cent of its entire programming output.[46]

While public broadcasting should not live or die merely according to the conventional ratings as determined by market share, SBS programs are generally not attracting sufficient audiences, although there are notable exceptions such as the big audiences it attracted during the World Cup soccer in 1998. SBS TV has not captured core audiences, only minority numbers in terms of total market share, yet it has had some solid numbers of committed viewers across programming genres. Nigel Milan, Managing Director, argues, 'I don't believe we judge ourselves on the ratings. I don't think that's what SBS is about. To fulfil our charter, we couldn't do that and be popularist. Our charter is about contributing to a cohesive and equitable inter-racial society.'[47] If Milan's justification of the role of SBS is accepted, then never before has the existence and vitality of a multicultural broadcasting organisation with the voice and vision of an SBS been more needed in Australian society.

There is a third tier of non-commercial broadcasting in Australia, namely community radio and television, sometimes referred to as *public broadcasting* rather than *public service broadcasting* because it has no consolidated institutional base paid for annually by government appropriation, like the ABC and SBS. Its origins lie in strong community demands to open up the media to wider constituencies. In the first 40 years of Australian broadcasting only two groups were allocated licences—commercial organisations and the ABC. By 1970, however, some 2000 requests for new broadcasting licences had been rejected, but the coincidence of a grass-roots radio movement and the election of the 'it's time' Whitlam government in 1972 saw the emergence of a new Australian broadcasting sector. Its advocates argue that community radio emerged as a *public broadcasting service* rather than as *public service broadcasting*. The sources of requests for new radio licences came from FM music enthusiasts, educational institutions, ethnic communities disenfranchised by the established Anglo-Saxon broadcasting monopolies, and other diverse groups who sought direct community access for participatory programming. Radio 5UV-AM, at the University of Adelaide, was Australia's first public radio station in 1972, followed by 3MBS-FM and 2MBS-FM in 1974, stations of the Music Broadcasting Society of Victoria and New South Wales respectively. In Melbourne in 1975, 3CR went to air as the first community radio station supported by a diverse group of community interests, including strong trade union support.

Contrary to fears at the time for the future of this fledgling alternative broadcasting sector under the Fraser administration elected in 1975, Minister Tony Staley gave this public broadcasting sector a secure legislative foundation. Under amendments to the *Broadcasting and Television Act 1976*, three types of public broadcasting stations were defined: category *E licences* for educational bodies; category *S licences* for groups intending to serve 'a particular interest or group of interests'; and category *C licences* to community groups 'intending to provide programs serving a particular community'. A 'scatty' but refreshing new sector of Australian broadcasting had been born, although without guaranteed funding from governments and advertising the viability of these stations was always going to be a problem. Lack of adequate income has meant that the energies of the few paid staff tend to be sapped by fund-raising drives in this highly responsive and participatory Australian broadcasting sector.

One might have expected to see the emergence of a promising community television sector during the period of five successive Labor governments (1983–96). After an extended period of lobbying by aspirant community television groups, the Keating government finally granted some open narrowcast licences to aspirant groups in 1993. However, these were established in such a way that inevitably the survival of community television would be threatened. There was no thinking by policy-makers to consolidate this small, but participatory and worthwhile innovation. Melbourne's Channel 31, for instance, was able to purchase its transmitter as a result of a grant in 1994 from the Victorian Harness Racing Board. This meant that the station had some funding, and it enabled trotting to be broadcast one night a week. The channel offered opportunities for novice on-air presenters and programs of specialist interest—community news and local district variety programs, live horse racing on Sundays, rap music, skateboarding and gay night life. Channel 31 boasted that it was 'the only station to bring suburban weather reports every hour!'. However, typically for community broadcasting, when its core funding was later lost (to pay television) the station found itself in dire financial trouble. The pattern is familiar for community broadcasters who were never established in a way which properly acknowledged their role in the broadcast system and genuinely supported their existence.

We have not been good at seeing how important it is to open up our media, to provide opportunities for far greater access to traditional media to many more people, and to use this era of change to provide much greater ideological diversity in our media. We must begin to think much more constructively about the great potential that public and community broadcasting can offer the nation in vexed times.

4 Citizens to customers: Telecommunications in transition

> *It was perhaps inevitable that the country of the emu and the platypus should design a bird with curiously clipped wings and then complain that it could not fly for profit. Such is the story of Aussat: a story lost not in the mists of time, but rather in a fog of unresolved policy issues. As the short history of Aussat is almost the complete story of competition in telecommunications in Australia.*
>
> Peter Leonard, 'Footprints down a narrow path',
> Media Information Australia, 1990

> *In Australia, as elsewhere, much time and effort was devoted to trying to delineate sharp boundaries between areas of open competition and those reserved to the carriers . . . The lesson is that in the longer run, the market, like love, laughs at locksmiths.*
>
> Henry Ergas, 'An alternative view of Australian telecommunications reform', Implementing Reforms in the Telecommunications Sector: Lessons from Experience, eds B. Wellenius and P. Stern, Aldershot, Avebury, 1996, p. 252

Alexander Graham Bell's patent of the telephone in 1876 contributed significantly to social and economic transformation in the twentieth century. The telephone became a major social instrument in most people's lives by providing companionship, flexibility and convenience in the way many people could communicate with each other. It was a key catalyst of the communications revolution as it became more intelligent in its applications and more mobile in its functionality. By the 1930s business had become more dependent upon the telephone, by the 1950s telecommunications networks had become global, and by the 1990s data came to challenge telephony as the leading mode of telecommunications. The telephone network and its companion technology, the computer, increasingly came to merge with each other to facilitate remarkable communications innovations, with the most notable child of this convergence, the Internet, reaching significant take-up in the 1990s. Telecommunications is widely predicted to become the world's biggest industry early in the new century.

Telecommunications has undergone profound technological and political changes, especially in recent years. Its history has been essentially one of monopoly public ownership around the world during decades of expensive and unprofitable public investment in building the networks. For much of the twentieth century until the mid-1980s, the institutional structure for the operation of telecommunications was similar in most countries—that of a legal monopoly with authority exercised directly or indirectly by the state. The consensus was that a public telecommunications authority (PTA) constituted a natural monopoly, like water supply, where, in theory, one firm could produce services at lower average costs than could two or more firms. The natural monopoly model was long considered as the only way of constructing national infrastructure and delivering telecommunications services to all. Given the higher relative costs of the construction of telecommunications services in rural than urban areas, the telecommunications industry was historically considered as inappropriate for major private sector ownership and control, on the grounds that unregulated profit-oriented firms would not provide universal public access to telephone services, or it would lead to systems that were operationally compatible.

The natural monopoly model was widely accepted until the late 1970s in the United Kingdom, in France and West Germany, in Sweden, Norway and the Netherlands, in Japan and throughout South-East Asia, and in New Zealand and Australia. The most notable exception was the United States, which built its system around heavily regulated private monopolies, notably American Telephone and Telegraph (AT&T), which grew out of the early Bell Telephone company to become the world's richest private monopoly until it was broken up in the mid-1980s. For Australia, of course, the development of telecommunications was always going to be of great national significance in a vast country so geographically isolated from the rest of the world. Australia had to find its own appropriate mix of affordable economic development of network infrastructure together with ways of providing access and social equity to such widely dispersed users.

A series of complex international political economy factors in the early to mid-1980s saw the ideology of international telecommunications shift towards growing acceptance of a more private corporate, deregulated and liberalised environment. The shift towards structural reform of the telecommunications industry, which began in the 1980s, gathered momentum as a worldwide wave of change. As major telecommunications carriers such as British Telecom, NTT (Japan) and Telecom New Zealand became privatised, telecommunications became a major focus of world investment. The General Agreement on Tariffs and Trade (GATT) and the World Trade Organization (WTO) were used as vehicles to liberalise international trading practices, and the industry's most important professional forum, the International Telecommunications Union (ITU), began to deal with complex technical solutions for the implementation of liberalisation. The changes in the Australian telecommunications industry can only be understood in the context of major forces driving changes in the international political economy.

An ideological transformation has been under way within many nation states since the late 1970s/early 1980s. The US economist Milton Friedman successfully promulgated his brand of economic theory called monetarism, arguing for the reduction of government spending, the freeing up of the rules for the private sector, a reduction in the role of public bureaucracy, and promotion of a more competitive, entrepreneurial culture. Britain's Prime Minister Margaret Thatcher was the quintessential applied economic rationalist, savagely cutting government spending in the 1980s, embarking on wholesale privatisation, including that of British Telecom in 1984, deconstructing the welfare state and disempowering Britain's trade union movement. Thatcherism became the economic paradigm of Western economies during the 1980s and 1990s. Developing economies also recognised the close connection between their future national prosperity and the level of advancement of their telecommunications system, and progressively encouraged private sector investment in both incumbent and new carriers.

These major changes to the ownership of telecommunications have also coincided with significant technological changes (outlined earlier in Chapter 2), especially digitalisation, dramatically increased network capacity and intelligence, and convergence. However, it is the human managerial face driving change, within governments, corporations and the financial sector, and the growth of more sophisticated consumer demand for services, which has driven the radical reorganisation of the world's telecommunications industry in recent years.

Some key forces at work driving the changes in the telecommunications industry may be briefly summarised as follows:

- Rejection of the acceptance of natural monopoly arguments grounded in what came to be seen as anachronistic assumptions about the

necessity and outcomes of public sector ownership of telecommunications.
- A fundamental re-evaluation of the role of the public sector, given increasing government deficits around the world and concern at the size, cost and efficiency of the public sector.
- Given the presumed nexus between an advanced telecommunications system and a strong economy, it was necessary to attract significant private investment and new entrants to create a dynamic telecommunications industry.
- The emergence of an interdependent global economy with liberalised trade policies, and the perceived necessity of joining 'the electronic superhighways club'.
- The convergence of media, information technology and telecommunications, which opened up extraordinary product development opportunities for information technology and telecommunications companies, and powerful new distribution channels for the major content providers.
- Major technological developments, especially digitalisation and dramatically increased network capacity and intelligence, leading to a proliferation of new communications services.
- Unprecedented consumer demand for telecommunications services, especially in mobile telephony and Internet-based services.

At the heart of major changes to telecommunications was a fundamental shift in the locus of power in society, away from one built on foundations of public ownership, public services and public interest, towards a more pluralistic, more privatised and more entrepreneurial/risk-taking culture. It is an industry which is growing exponentially and, at present, is in a phase of radically changing its identity.

A network for the nation

In a vast but sparsely populated country, Australia's telecommunications history has been driven by the national goal of overcoming vast distances and breaking down isolation. Australia is one of the most urbanised countries in the world: a continent of two million square kilometres yet with about 80 per cent of its total population living in five cities—namely Sydney, Melbourne, Brisbane, Adelaide and Perth. The benchmark policy principle, to provide universal service at affordable prices, has resulted in Australia having one of the highest telephone densities per capita in the world, with tariffs reasonably priced by world comparison, and with ample connection to the rest of the world.

Australians have long shown a propensity to quickly adopt communications technologies. In 1854, Australia's initial telegraph service clicked its Morse code messages over the nineteen kilometres between Melbourne and Williamstown. By 1858, poles with single iron wires linked Melbourne, Sydney and Adelaide. A year later, the remarkable achievement of linking Melbourne to the island of Tasmania by submarine cables across Bass Strait was completed. In the 1870s, Charles Todd undertook to construct a 2900-kilometre overland telegraph linking Australia top to bottom, from Adelaide to Darwin. Using information chronicled by the explorer John Stuart, who had made the first crossing of the continent just eight years earlier, traversing one of the harshest environments in the world, Todd completed the project in 1872. Australia boasts a rich history of extraordinary telecommunications pioneers.[1]

When the Australian states agreed to a federation in 1900, the Constitution vested the responsibilities for postal, telegraphic and telephonic services with the national government. With Federation in 1901 came the formation of the Postmaster-General's Department which was granted authority to 'establish, erect, maintain, and use stations and appliances for the purpose of transmission and receipt of wireless messages'. Ann Moyal's classic history of Australia's telecommunications, *Clear Across Australia*,[2] details the early problems of harnessing the disparate postal, telegraph and telephone services into a central administration, and she reminds us that the most vexed issues in Australia's communications policies—dissatisfaction with services, cross-subsidies for rural services, interstate rivalries—are as old as the Commonwealth itself.

In 1960, significant technical changes fashioned modern telephony, including the introduction of cross-bar exchanges, supporting subscriber-initiated automatic trunk dialling. The policy foundation was the Community Telephone Plan (1960) of the Postmaster-General's Department developed by the Australian Post Office, which was then responsible for both postal and telecommunications services, to progressively improve new connections service, increase the number of automatic exchanges, improve distant transmission standards and enable telephone users to dial any other subscriber within Australia. Governments after the Second World War, and until the early 1970s, were a Coalition of two conservative parties, the Liberal Party and the Country (now National) Party. The Country Party naturally showed special interest in communications on behalf of its rural constituents, and it usually held the Postmaster- General's portfolio in government.

The election of the first Labor government after 23 years in Opposition, swept to power in 1972 on the theme of 'it's time for change', heralded many institutional changes. In keeping with international trends at that time, responsibilities for postal and telecommunications were divided between an Australian Postal Commission (Australia Post) and Telecom Australia, established as a statutory authority under the *Telecommunications Act 1975*. At that time, responsibility for overseas telecommunications became a politically contentious issue. The Overseas Telecommunications Commission (OTC) had been established in 1946 as a Commonwealth business enterprise by amalgamation of a Cable and Wireless subsidiary and Amalgamated Wireless Australia Ltd. The Whitlam Labor government (1972–75) pushed for the merger of OTC at the time with the newly created Telecom, but a hostile Senate Upper House voted against this. It is one of the grand institutional ironies of Australia's telecommunications history that in 1990, when a Labor government decided to sell Aussat, the domestic satellite company, Cable and Wireless, whose Australian subsidiary was bought out by an earlier Labor government in 1946, was to buy 49 per cent of the successful competitive tenderer, Optus Communications.

Aussat became an institutional player in Australian telecommunications in the early 1980s. The catalyst for the introduction of a domestic satellite system came from Channel 9's commercial television network proprietor, Kerry Packer. Packer knew that a domestic satellite system offered him the prospect of increased television audiences in rural Australia, but also the potential to break Telecom's domestic telecommunications monopoly. Packer's request, as is often the case in Australia's communications, became national policy in 1979. Tony Staley, then Minister of Posts and Telecommunications, justified the decision to establish a domestic satellite system in a classic 'tyranny of distance' speech when he told the House, 'It is too easy to overlook, or remain blissfully oblivious to, the plight of our fellow countrymen who are seriously disadvantaged by a lack of communications services and communications dependent services.' The then Liberal Coalition government proposed Aussat as a commercial company, though with only 49 per cent private ownership. When the Hawke administration with its strong trade union base came to power in 1983, 75 per cent of the shares were held by the Commonwealth and 25 per cent by Telecom Australia. Given the huge infrastructure costs of setting up a domestic satellite company, and the fact that the market was heavily regulated in favour of Telecom and OTC, Aussat struggled to build a viable business.

In May 1988, Gareth Evans, then Minister for Transport and Communications, tabled *Australian Telecommunications Services: A New Framework* which advocated a gradual shift towards deregulation of the Australian telecommunications market. It may be the only time in Australian telecommunications policy history that a government outlined its objectives for the industry: These were to:

a. Ensure universal access to standard telephone services throughout Australia on an equitable basis and at affordable prices, in recognition of the social importance of these services.
b. Maximise the efficiency of the publicly owned telecommunications enterprises—Telecom, OTC and Aussat—in meeting their objectives, including fulfilment of specific community service obligations and the generation of appropriate returns on investment.
c. Ensure the highest possible levels of accountability and responsiveness to customer and community needs on the part of the telecommunications enterprises.
d. Provide the capacity to achieve optimal rates of expansion and modernisation of the telecommunications system, including the introduction of new and diverse services.
e. Enable all elements of the Australian telecommunications system (manufacturing, services, information provision) to participate effectively in the rapidly growing Australian and world telecommunications market.
f. Promote the development of other sectors of the economy through the commercial provision of a full range of modern telecommunications services at the lowest possible costs.

Labor was flagging the winds of change, although by international standards at the time this was incremental in terms of policy change and politically cautious.

The most innovative aspect of the Evans policy document was the proposal to establish a new telecommunications regulatory authority called Austel. There had long been complaints that in Australian telecommunications, Telecom had been principal player, umpire and arbitrator, and many advocates proposed the need for an independent regulator, a principle with which Telecom itself did not disagree. Austel was to be a single specialised telecommunications regulatory agency, independent of the carriers and answerable to the government through the Minister of Communications. Its principal responsibilities were to maintain technical standards, monitor fair and efficient market conduct, including possible breaches of the *Trade Practices Act*, and protect consumers, especially from monopoly price control.

Inevitably, Australian telecommunications was to be swept up in the tide of deregulation that was gathering international momentum. By the late 1980s the several sources advocating change of the Australian

Table 4.1 Australian Telecommunications carriers, 1991

	Telecom	OTC	Aussat
Established	1975	1946	1981
Revenue (1991)	$9531.2m	$1891.2m	$158.4m
Pre-tax profit	$1625.5m	$453.2m	–$15.1m
Post-tax profit	$962.7m	$275.3m	–$15.1m
Workforce	84 000 (1990/91)	21 000 (1989/90)	346 (1990/91)
AOTC revenue (1991)	$11 224.4m	–	–
Post-tax profit	$1048.0m	–	–

Source: Annual Reports, 1991–92.

telecommunications system were gathering momentum, especially pressures for a more competitive regime, and key private corporate players and merchant bankers who wanted a share of this increasingly lucrative market through privatisation. The Hawke Labor government faced a pressing telecommunications industry problem in 1990—Aussat's continuing poor trading performance and the company's debt. Table 4.1 shows the respective trading position of the three carriers in 1991, including Aussat's relatively weak financial position.

The Labor government canvassed institutional change in an attempt to solve the Aussat debt problem. Telecom understandably wanted to maintain its prime market position, although it was aware that the political climate demanded change. Telecom called for the government to give it a much freer hand in terms of management accountability. Its then Managing Director, Mel Ward, called for the abolition of restrictions on levels of capital borrowing, on conditions for the employment of staff and reductions in purchasing policy, arguing that it must be allowed 'to manage its business and to be accountable to its owners and its shareholders on results'.[3] Not quite as ostrich-like as was often alleged at that time, Telecom actually proposed in the 1990 institutional review the notion that it compete with a foreign second carrier. The other major vested interest positions were that, in summary, certain major business interests wanted a greater share of the growth, arguing for a substantial diminution of Telecom power and a more competitive regime; trade union interests, with considerable political clout, wanted to ensure that the industrial status quo would be maintained; and equipment suppliers, mostly subsidiaries of transnational corporations, notably Ericsson and Alcatel, wanted the continuation of Telecom's long-term preferential contracts.

In November 1990 the Australian government announced its intention to introduce wide-ranging changes to the Australian telecommunications environment. Aussat, the debt-ridden satellite carrier, was to be sold 'at a price determined by tender with a licence to provide a full range of domestic and international services'—an unprecedented move by the Australian Labor Party into privatisation in telecommunications. This was a period more commonly known as the *Telstra/Optus duopoly*, with these two carriers licensed to supply a full range of domestic and international services, over both fixed and mobile networks. The duopoly consisted of a merged Telecom/OTC, temporarily named the Australian and Overseas Telecommunications Corporation (AOTC) but renamed Telstra in 1993, and Optus, which bought satellite operator Aussat's assets in return for its telecommunications licences. Optus was a private consortium, then 49 per cent owned by Bell South (US) and Cable and Wireless (UK)—each with 24.5 per cent at the time of the formation of the duopoly, together with a 51 per cent Australian shareholding, including security company Mayne Nickless as a 25 per cent shareholder. At the time the licence was valued as high as $4.5 billion, but the successful tenderer, Optus, paid $850 million, a figure it later considered too high given its continued difficulty in trading profitably during the period 1992–97. In August 1997 the Howard government removed the foreign investment restrictions on Optus, enabling Cable and Wireless to buy the Bell South shares, thereby gaining control of the company with a 52 per cent stakeholding. The incumbent Telstra, and Optus, a 'greenfields' company, were legislated as duopolist telecommunications carriers until 30 June 1997. In mobile telephones, on the basis of anticipated growth of the market, the government later granted a third digital mobile carrier licence, to Vodafone.

The comparative trading pattern of the telecommunications duopolists from 1992/93 to 1996/97, the nation's fastest growth industry, is shown in Table 4.2.

Optus was always destined to face difficult issues of strategic and financial development. Initially it was, understandably, highly dependent on the Telecom network which had been progressively publicly financed and built during eight decades of the twentieth century. Under its licence conditions, Optus was obliged to provide its own infrastructure to connect all Australian capital cities. In March 1993 the first telephone calls were made from Optus's own directly connected network; in May 1993 it launched Optus Mobile Digital; and in September 1994 the company established its pay television company, Optus Vision. The staggering cost of network investments, the most significant of any private sector company in Australia during the 1990s, put enormous pressure on its balance sheet, making its principal stakeholders anxious about the overall trading performance and their return on capital invested in Optus. These tensions are reflected in their three changes

Table 4.2 Australian telecommunications carrier revenue/profit–loss, 1992/93 to 1997/98 (Total sales operating revenue, and profit)

	Telstra ($m)	Optus ($m)
1993		
Revenue	12 656	–
Profit (after tax)	905	–
1994		
Revenue	13 363	841.3
Profit	1 699	(97.7)
1995		
Revenue	14 081	1 434.5
Profit	1 753	(17.0)
1996		
Revenue	15 239	1 994.0
Profit	2 305	60.3
1997		
Revenue	15 983	2 505.6
Profit	1 617	(411.8)
1998		
Revenue	17 302	2 933.3
Profit	3 004	(95.4)

Sources: Telstra Annual Reports; and Optus Financial Data Book, April 1998.

of CEO in such a short period—initially Bob Mansfield from McDonald's, then Dr Ziggy Switkowski from Kodak, followed by Chris Anderson from Fairfax. In some of the most extraordinary appointments in Australian senior personnel history, Dr Switkowski later became the CEO of Telstra, replacing the retired Frank Blount in February 1999; Bob Mansfield later followed as Chairman of Telstra.

In November 1998, Cable and Wireless Optus Ltd made a stunning stock market debut; retail investors who had paid $1.85 and institutional shareholders who had paid $2.15 a share were delighted to see the shares reaching $2.74 on the first trading day. Chief Executive Chris Anderson declared that the company was poised to make its maiden profit and added, '[W]e are a very strong company—$3.5 billion worth of turnover, a 17% growth rate and we took $350 million out of costs last year.'[4] However, the general investor consultant advice was that Cable and Wireless Optus could not be expected to pay a dividend for another two or three years, and that it was a much riskier stock than Telstra.

Optus management has long been aggravated by Telstra's continuing command of the marketplace, annoyed that the promised fruits of deregulation have taken so long to appear, and frustrated by delays in the regulatory process.

The seeds were sown for a later shift to a model of open competition, and the *Telecommunications Act 1991* actually enacted a sunset clause for the end of the duopoly on 1 July 1997. For seven decades the *raison d'être* of Australian telecommunications had been to build a national network to provide universal service over the vast distances of Australia with substantial public investment. Telecommunications ran at large losses for many decades, but the outcome was that Australia has one of the best national telecommunications networks in the world, which has been transformed into an immensely profitable and strategically important business in the late 1990s.

Table 4.3 Australian licensed telecommunications carriers, post-1 July 1997 (in order of licensing, as at July 1998, with the latest to be registered last)

Carrier licence granted to	Date licence granted
Telstra Corporation Ltd	1 July 1997
Optus Networks Pty Ltd	1 July 1997
Optus Mobile Pty Ltd	1 July 1997
Vodafone Pty Ltd	1 July 1997
AAP Telecommunications Pty Ltd	1 July 1997
Primus Telecommunications Pty Ltd	1 July 1997
Optus Vision Pty Ltd	1 July 1997
Telstra Multimedia Pty Ltd	1 July 1997
Horizon Telecommunications Pty Ltd	25 July 1997
OMNIconnect Pty Ltd	19 August 1997
United Energy Telecommunications Pty Ltd	27 August 1997
Windytide Pty Ltd	4 September 1997
Northgate Communications Australia, Ballarat Pty Ltd	3 December 1997
Macrocom Pty Ltd	18 December 1997
Oz Telecom Pty Ltd	2 March 1998
WorldCom Australia Pty Ltd	24 March 1998
Iridium South Pacific Pty Ltd	2 April 1998
PanAmSat Asia Carrier Services Inc.	1 May 1998
POWERTEL Ltd (formerly known as Spectrum Network Systems Ltd)	6 May 1998
Agile Pty Ltd	15 May 1998
Xinhua News Telecommunications Pty Ltd	1 June 1998
Amcom Pty Ltd	28 July 1998
Davnet Pty Ltd	1 September 1998
SCCL Australia Ltd	22 September 1998
Hutchison Telecommunications (Australia) Ltd	30 September 1998
TransAct Carrier Pty Ltd	26 February 1999
Soul Pattinson Telecommunications Pty Ltd	19 March 1999
One.Tel GSM 1800 Pty Ltd	25 March 1999
Commcord Pty Ltd	29 June 1999
Communication Site Rentals Pty Ltd	1 July 1999
Wideband Access Pty Ltd	6 July 1999

Source: Australian Communications Authority, *http://www.aca.gov.au/licence/carrier/carriers.htm*, accessed on 30 August 1999.

The era of open competition and liberalisation, post-1 July 1997 was a fundamental policy change intended to open up the Australian telecommunications industry to newcomers in a more liberalised and highly competitive environment. Telstra, as entrenched incumbent, and Optus, protected duopolist for six years, obviously had significant head starts on their new competitors. The spirit of the legislation was to provide a competitive free-for-all, with no limitations on the number of telecommunications carriers (although they had to be licensed) or service providers, and with no distinction between types of carriers or carriage service providers. By July 1999 there were 31 licensed carriers in the new Australian telecommunications marketplace, as shown in Table 4.3.

Conflicting interests, conflicting objectives

The major policy changes in Australian telecommunications in the 1990s did not emerge as a result of clear blueprints for reform in the national interest. Rather, new policies have been the uneasy outcomes of serious conflicts in policy objectives, with governments weaving their way through thickets of vested interests, trying to balance political pragmatism with their sense of the need for reform. Policy conflicts have arisen between the macroeconomic and microeconomic goals of the government of the day (or any of the major political parties). There remain a series of complex conflicts between economic and social policy in Australian telecommunications. The tensions which have arisen between policy objectives may be briefly summarised as:

- generic economic competition policies vs specific industry sector policies;
- national goals vs industry goals;
- government intervention vs marketplace decision-making; and
- international policies vs local manufacturing.

Generic economic competition policies vs specific industry sector policies

The government had to decide in its reform process how quickly it could move from monopoly to the managed competition of a duopoly, then later to open competition. Its fear was that if it moved too quickly towards a competitive free-for-all, and the industry became destabilised, the political consequences would be serious. So, on the one hand the government wanted a much more competitive Australian industry marketplace with minimal government regulatory involvement, yet on the other hand it wanted to protect the viability of individual telecommunications industry competitors, particularly new entrants. Obviously an incumbent carrier comes to any new era of competition with long-established competitive advantages, especially in its control over the customer access network. To facilitate some immediate gains from competition policy, the government mandated that new entrants should be able to interconnect into the former monopolist's network at a fair price. However, setting the interconnection price required a difficult judgment to be made between Telstra, who understandably want to charge the highest figure possible for competitors to connect to its network, and the new entrants who want low network access fees to give them a competitive kick-start. One key dimension in assessing the long-term success of competition policy will be the number and viability of new entrants who come to the industry because of fair and effective interconnection policy.

National goals vs industry goals

Every Australian government since the early 1980s has faced the serious problem of high levels (officially between 8 per cent and 11 per cent) of registered unemployment, alarmingly, with over one million people out of work in the 1990s. Both the Labor and Coalition governments have had to grapple with the issue of whether they need to invest public funds into job creation schemes, while preferring job creation to emerge as an outcome of greater industrial productivity and efficiency. There is little evidence to support the general proposition that economic growth in telecommunications will lead to a substantial reduction in the level of unemployment. The telecommunications industry has shed thousands of jobs in Australia in order for it to become more internationally competitive. The creation of new jobs by new carriers and re-sellers during this growth phase for the industry has not fully offset the 25 000 staff that Telstra has shed during the 1990s, which always carries with it serious human disruption and national political consequences. So many talented senior staff have left Telstra in recent years that there is a sense that the company has lost its corporate memory. For most of the decades of network building during the twentieth century the telecommunications industry was seen as a great vehicle for employment growth in Australia. Not any more. The last decade has seen the widespread acceptance of substantial job losses for the sake of achieving greater international competitiveness. This is rarely analysed and is usually expressed as the 'pain we need to get gain'. The microeconomic reform objectives, driving for a more competitive Australian industry, compete with the vital national social goal of maintaining high levels of employment and people's sense of self-worth.

Government intervention vs marketplace decision-making

The relationship between Telstra and the government has long had many tensions within it. As the level of public debt increased from the 1980s, the government needed to rein in overall public expenditure and hence control the public carrier's level of borrowing. Yet, in a telecommunications marketplace which has become progressively more competitive, investment decisions, including high-risk strategic decisions, surely ought to be largely left to the company management itself. The government wants to promote competition and market-based decision-making on the one hand, yet it still keeps major responsibilities in so many regulatory matters, such as interconnection arrangements, a whole range of pricing decisions and the rules for universal service obligations. One of the greatest ironies of the age of deregulation in telecommunications around the world is that it is also an era of the proliferation of regulatory bodies, with increasingly complex regulations.

Privatisation presents classic dilemmas for both parties. Any government which decides to progressively privatise its principal carrier, as has the Howard government in the late 1990s, needs it to maintain excellent profits in order to attract more investors to the next stage of the float. Hence, one arm of policy works to inflate the consumer prices to ensure good company profits for the maximum gain from the privatisation of Telstra, yet the other arm of policy is promising that competition will deliver huge reductions in telephone charges to consumers. These policy objectives are essentially irreconcilable, placing the carrier management in an impossible bind. Since there are no clearly articulated long-term national objectives in Australian telecommunications, the carrier finds itself adapting to the whims of short-term politically expedient pressures.

International policies vs local manufacturing

A vexed area for any Australian government is that of industry development policy. What responsibilities does government have, if any, to foster the development of local industry, including local manufacturing? Our history of industry development has been one of considerable government policy involvement.

Briefly, soon after the Second World War, members of the Australian Parliament expressed their concern at the high level of imports of telecommunications equipment. It was agreed that the purchasing leverage of the Postmaster-General's Department ought to be used to 'pump prime' the local electronics industry, using competitive tendering for telecommunications equipment to advantage local manufacture. This public policy was a major contributing factor that brought a group of transnational corporations to set up world-class manufacturing plants in Australia: Ericsson from Sweden, Alcatel from France, Siemens from Germany, and later NEC and Fujitsu from Japan, and Nortel from Canada. An industry development model emerged of a publicly owned telecommunications network infrastructure served by a series of privately owned companies, mainly transnationals, but also the Australian-owned AWA, Datacraft and JNA, part of an electronics manufacturing industry which overall employed about 10 000 people in Australia in the 1970s and early 1980s.

Transnational corporations work within the rules of each country, and this policy of using the purchasing leverage of the common carrier had considerable multiplier effects. Later again, in 1987, under the Button information industry plan, nearly all of the transnationals signed partnership agreements with government to build more exports in telecommunications from Australia and to increase their level of research and development (R&D) investment and local manufacturing capability in Australia. The policy was an attempt at reducing the substantial imbalance between Australia's imports and exports of telecommunications goods and services, as well as encouraging technology transfer to build small and medium-sized Australian enterprises. One estimate of the national benefit is that the partnership for development plan, together with other industry development initiatives, had export commitments by 1995 (actual and forecast) of $8700 million, together with R&D commitments of $2300 million.[5] Although essentially foreign-owned, Australia established a world-class electronics industry in telecommunications manufacturing through the cumulative effects of these industry development policies.

So, how does the new economic paradigm for telecommunications, which centres on global suppliers rather than local suppliers, work for Australia? Is it still important to promote local industry development, retain significant local manufacturing, and devise policies for small and medium-sized Australian enterprises to grow? Can we actually go global but still think locally in this new industry policy climate?

How have Australian governments and the telecommunications industry responded to or resolved these many conflicts of policy interest? What were the landmark decisions for the open competition era? In summary, the critical policy issues in Australia's new telecommunications policy regime are competition policy, interconnection for new players, pricing policy, and the issues surrounding universal service obligations. Each of these will now be analysed.

Deregulation: The awesome foursome

Competition policy

A key factor in the growth of the international telecommunications industry has been the level of international cooperation between public carriers, best exemplified by the post-Second World War establishment of Intelsat, an international body established to develop and manage the rules governing world satellites which has operated with remarkable cooperation over 40 years among dozens of nations. Also, the collaborative development of technical standards by the major carriers and their vendors at fora convened by the International Telecommunications Union (ITU) ensured the compatibility of services now taken for granted, such as international direct dial (IDD), modems and fax services. This ethos of international cross-industry cooperation represents an essential difference between the history of the telecommunications industry as opposed to that of the generally more maverick, self-interested computer industry. Today, however, the driving forces of the telecommunications industry are increasingly like the computer industry, primarily centred on economic assumptions about efficiency, competition, privatisation and deregulated markets.

Australian telecommunications policy has a long history of government intervention in matters such as infrastructure investment, pricing of services, borrowing levels by the common carrier, purchasing practices, carrier staffing levels, the possible introduction of new technologies such as pay television, and industry development policy. However, unlike the broadcasting industry, telecommunications during the 1960s to 1980s did not have the equivalent of an Australian Broadcasting Tribunal as a designated regulatory body for the industry until Austel was created as its first industry-specific regulator in 1989. All industry was subject to Australia's competition policy, enshrined in the *Trade Practices Act 1974*, which was primarily designed to prohibit anti-competitive practices, although it did provide anti-competitive exemptions provisions. Policy in the 1990s assumed that the Australian economy would be better served by less government intervention and more market-based approaches. Is it best practice for all competition in telecommunications to be regulated under a general competition law, or should a degree of industry-specific regulation for the telecommunications industry remain?

The seminal policy document on competition policy for the Australian economy was the outcome of the 1993 Hilmer Committee's inquiry, *A National Competition Policy for Australia*, which concluded that 'the greatest impediment to enhanced competition in many key areas of the economy are the restrictions imposed through government regulation'.[6] Hilmer proposed a uniform approach across all Australian government-owned and operated organisations, which he advocated should receive the same treatment as private companies. In a rare manifestation of political unity, these principles were endorsed by Australia's federal, state and territory governments in April 1995. All governments agreed to structural reform of public enterprises, limits on monopolistic pricing, and the eventual harmonising of their legal frameworks to promote competition. Many of the initial reforms which emerged as a result of Hilmer's work were in state-based enterprises, such as electricity, gas, water and transport infrastructure, and these experiences have had an important bearing on his thinking about the telecommunications industry. The notion that certain industries have specific development needs which ought to be given particular policy consideration was not accepted by Hilmer. The position taken was that if these other network-based utilities can be regulated without industry-specific rules, why should the telecommunications industry have one? Why was it different?

Hilmer's recommendation for a generic competition regulator was implemented, and the Australian Competition and Consumer Commission became the overall industry competition watchdog, overtaking the role of Austel from 1 July 1997, and relying on the new Parts XIB and XIC of the *Trade Practices Act* to oversee anti-competitive practices specific to the access regime in the telecommunications industry. ACCC decisions are open to appeal through the Australian Competition Tribunal and the Federal Court. What had been intended as a review and refinement of the telecommunications regulatory framework in 1991 shifted to generic industry regulation for competition and access by the time of the 1997 legislation. Austel's modified successor, the Australian Communications Authority (ACA), was to be responsible for technical regulation and industry organisational matters. The notion of fostering and building a real understanding of a key strategic industry through the expertise of a specific industry regulator was abandoned—yet another policy triumph for Canberra's economic rationalists.

Against this backdrop the *Telecommunications Act 1997*, together with associated bills, passed the Senate on 24 March 1997, just in time to meet the prescribed deadline to dismantle the Telstra–Optus duopoly by 1 July 1997. Maintaining the spirit of Hilmerism, other significant features of the new regime were as follows:

- There was no limitation on the number of telecommunications carriers or service providers in the industry. As had been intended in setting the sunset clause to the *Telecommunications Act 1991*, as from 1 July 1997, the carrier duopoly was generalised to an unlimited multi-carrier environment.
- There was no distinction by way of regulatory exclusiveness between the types of carriers (for example, fixed or mobile) or carriage service providers. Any carrier or service provider could provide any telecommunications service in Australia and offshore. The distinction between 'carrier' and 'carriage service provider' was changed to that of infrastructure ownership, rather than the right to provide certain network services.
- Although all carriers must be licensed by the ACA, carriage service providers were not.
- No technical, financial or roll-out requirements were to be imposed on carriers.
- The ACCC would be the economic regulator administering competition policy provisions, and the ACA—a merger of Austel and the Spectrum Management Authority (SMA)—would be the technical regulator, with virtually no scope for independent initiatives unless directed by the ACCC or the Minister.

The responsibilities for Australian telecommunications under the revised model are summarised in Figure 4.1.

Hence, the spirit of this legislation was to 'let a thousand flowers bloom', leaving the players to set the new rules among themselves in an open marketplace, but enabling them to go to the fledgling ACCC arbiter to sort out any disputes where commercial negotiation failed. The ACCC was given real teeth as the new regulator, notably the ability to apply demarcation fines if its competition articles were ignored. On 28 May 1998 the ACCC issued its first Competition Notice alleging that Telstra had engaged in anti-competitive conduct in breach of the telecommunications provisions of the *Trade Practices Act*. The notice alleged that Telstra contravened the competition rule by charging its Internet customers for services by Telstra while at the same time refusing to pay for a similar service it received from those same Internet customers. Then, on 10 August 1998, Telstra faced a $10 million-plus penalty after the ACCC issued a Competition Notice in relation to its customer transfer process which stated that Telstra was engaging in anti-competitive

Figure 4.1 The regulation of telecommunications in Australia

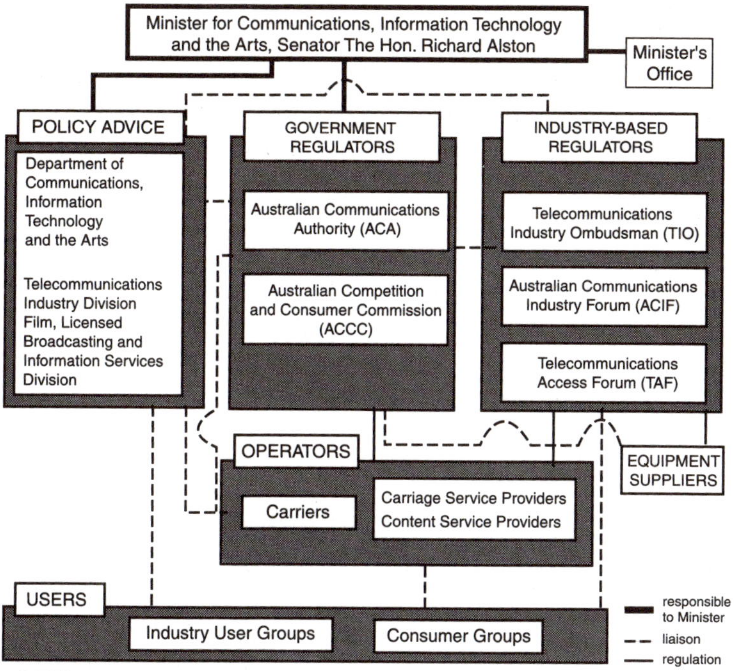

```
                    Minister for Communications, Information Technology
                    and the Arts, Senator The Hon. Richard Alston          ─── Minister's Office

      POLICY ADVICE              GOVERNMENT              INDUSTRY-BASED
                                 REGULATORS              REGULATORS
      Department of
      Communications,
      Information                Australian Communications   Telecommunications
      Technology                 Authority (ACA)             Industry Ombudsman (TIO)
      and the Arts

      Telecommunications         Australian Competition      Australian Communications
      Industry Division          and Consumer Commission     Industry Forum (ACIF)
      Film, Licensed             (ACCC)
      Broadcasting and
      Information Services                                   Telecommunications
      Division                                               Access Forum (TAF)

                      OPERATORS                                    EQUIPMENT
                                  Carriage Service Providers       SUPPLIERS
                      Carriers    Content Service Providers

      USERS
                Industry User Groups    Consumer Groups      ─── responsible to Minister
                                                             --- liaison
                                                             ─── regulation
```

Source: DCITA 1998.

conduct. Named as 'victims' were the carriage service providers AAPT, Macquarie Corporate Telecommunications, Optus and Switch Communications. The sticking point was the fee of $30 charged by Telstra when it transfers them to the accounts of local customers.

This really is a new era for Australian telecommunications, the history of which had previously been that of a strong public planning ethos, a dominant carrier and industry-specific regulation through Austel in the early 1990s. With the coming of the ACCC the Australian policy model has shifted more towards the American way of doing things, and with prime judgments emanating from the regulator and the courts rather than from a federal government—by design.

Interconnection

Interconnection is a remarkable phenomenon of competition policy in that a new entrant to the telecommunications industry can have access to declared elements of the incumbent's (or other) established network. This can be done by commercial negotiation between carriers, or by mandated access. Hence, for example, when interconnection began, an

Optus long-distance phone call may have been originated and terminated through Telstra's network, with the customer billed by the competitor, Optus, which subsequently repaid Telstra the agreed interconnection fee for the call. The policy rationale was that if a new player, such as Optus, could not have access on reasonable terms to Telstra's network, then the long lead time for the new competitor to establish its own network could jeopardise its commercial viability and the benefits of competition might be delayed, or lost entirely. Hence, interconnection refers to the access arrangements, for technical interoperability and pricing, between the incumbent carrier (or carriers) and new entrants who may not own infrastructure. Telecommunications is different from other utility-based industries in that interconnection is absolutely essential to permit end-to-end services in a multi-operator environment.

Jim Holmes has made two major points about interconnection:

1. The regulatory arrangements for mandated interconnection are most important in the early stages of network competition when new entrants have everything to gain and the incumbent has everything to lose. There is nothing in it for the incumbent, so delay is valuable.
2. The experience and confidence of the regulator needs to be strongest at the outset of competition, because that is when the regulator is most needed. Ironically, this is the time when the regulator is building experience and is testing the new legal framework.[7]

The *Telecommunications Act 1991* provided for Telstra and Optus to negotiate interconnection charges, but if the carriers could not agree on the rate then Austel was empowered to intervene. In June 1991 Austel was required to directly determine the initial interconnection charge, which it fixed at 3.14 cents per minute, with the charge incurred at both ends of the call. So, where one Telstra-connected customer called another Telstra-connected customer in Optus's long-distance network, Telstra charged Optus around 6.28 cents per minute of talking time. This interconnection charge was renegotiated in July 1994 at 3.5 cents per minute, with Austel acting as mediator this time rather than fixing the charge directly as it did previously. Telstra, of course, took the position that this was a generous interconnection figure given its huge network investment costs. Optus, however, could see little point in entering the local call market in a country with untimed local calls where it had to pay 6.28–7.00 cents a minute for every one of its customers' calls. Hence, once the call entered the fourth minute on the basis of a 25 cent charge, Optus was paying more in interconnection fees to Telstra than it could charge the consumer. Telstra's long-distance interconnection charges comprised 40 per cent of Optus's revenues for STD and IDD services. Hence, for new entrants there are complex trade-offs in terms of cost–benefit decisions in paying the interconnection access charges and being beholden to a competitor to provide network access, as opposed to the substantial capital investment in rolling out independently owned and controlled access networks. In practice, Optus paid both the high costs of rolling out its own long-distance and mobile network as well as substantial interconnection charges to its arch competitor.

An interesting regulatory contrast is provided by the *laissez-faire* New Zealand telecommunications policy model, where it took two years for incumbent carrier New Zealand Telecom and the new entrant, Clear Communications, to complete an interconnection agreement by May 1991. The counterpart 1994 agreement between Telecom Australia and Optus, under the leadership of regulator Austel, took only three months. There is a clear trend internationally where lightly regulated competition regimes tend to generate a telecommunications market which becomes increasingly litigious and costly for the stakeholders.

Once a company has been declared a carrier, the following access arrangements apply:

- A carrier must provide interconnection to the declared elements of its networks to other carriers. To date, interconnection agreements have centred on access to Telstra's network by new competitors, although they can also apply to a new competitor seeking access to the Optus network and apply to other new carriers as well.
- A carrier must provide adequate call data to other carriers to support access to declared services. With interconnection arrangements, the cards tend to be stacked in favour of the incumbent who knows, or presumably knows, the full financial and cost details of their operation. There is no requirement for the parties to make public the outcome of their commercial arrangements, although the ACCC can request such cost information. Holmes argues that 'transparency is a problem, and has been made secondary to the principle of commercial confidentiality. The result is that competitors are disadvantaged, and must proceed to their own negotiated outcomes in the dark.'[8]

From late 1998 the ACCC announced some major decisions about interconnection. On 14 December 1998 the ACCC issued a draft declaration that would require Telstra to allow its competitors direct access to the lines that link customers to local call exchanges. The rationale behind this declaration is to increase competition in the important local call market. On 19 January 1999 a draft ACCC determination rejected Telstra's proposed interconnection charges and concluded that the proposed charge of around 4 cents a minute by Telstra would need to be halved to be acceptable. For Telstra, such a change would cost them an estimated $200 million in lost revenue, but for major competitors such as Cable and Wireless Optus and AAP Telecommunications, interconnection charges make up about 35 per cent of their total costs for long-distance calls. ACCC Chairman Allan Fels was quoted as saying that 'halving interconnect charges could reduce the prices of national long-distance calls by up to 15 per cent'. Graeme Ward, Telstra's group director for regulatory and external affairs, responded that 'this draft determination leaves little incentive for Telstra to invest in its networks, which in the long run is detrimental for Australia, the industry and its consumers'.[9] The rules of interconnection will be a major battleground of competition policy during the next few years.

Pricing

One of the most vexed policy areas in telecommunications is pricing, especially how prices should be determined, by whom, and how prices should be regulated, if at all. In the days of Telecom's terrestrial monopoly, the prime policy objective of providing universal services at affordable prices inevitably meant that Telecom cross-subsidised its services. Section 6 of the *Telecommunications Act 1975* required Telecom to 'perform its functions in such a manner as will best meet the social, industrial and commercial needs of the Australian people and make its telecommunications services available throughout Australia for all people who reasonably require those services'. Hence this 1975 Act essentially called for Telecom, a publicly owned corporation, to operate within the framework of a social charter. Telecom had to cross-subsidise its operations; the super profits from the STD calls between Melbourne and Sydney could be used to provide phone services to outback Oodnadatta, for instance, which each cost about $20 000 to install but returned roughly only $400 per annum in telephone charges. If, of course, Telecom had charged the Oodnadatta consumer the real price of providing the service, then the telephone would have been unaffordable.

Politically, Australia's National or Country Party has been important in forming an alliance with the Liberal Party in order to be able to form a government, and Country Party members have long been effective at protecting the interests of their rural constituents with infrastructure investments and pricing equity. Hence, Australian telecommunications has a long history of 'skewed pricing', where Telecom ran a complex series of cross-subsidies across its entire operation. With the capture of the policy process by economic rationalists, and the consequent drive for competition and efficiency, the pressure has been for Telstra to dismantle its cross-subsidies, to relate its charging to real costs and to become more open about its costing data.

One of the best exercises in analysis of Telecom's pricing practices was provided by a committee of inquiry into telecommunications services appointed by the Fraser government in 1981 and headed by Jim Davidson. This report highlighted anomalies in cross-subsidies, and argued that uniform pricing led to the misallocation of funds, in that the financial burden for support of socially desirable objectives was borne by a select group of customers, instead of being spread over the whole community. Davidson challenged the pricing premise that low telephone rentals were provided for low-income families. For example, argued Davidson, price differentiation between business and residential telephone rentals existed with the effect that wealthy families enjoyed the same low-income subsidies as non-business users, while 'struggling businesses who provide employment pay discriminatory high telephone calls'.[10] Conversely, low-income families of metropolitan Australia were subsidising rural Australians, including wealthy rural subscribers.

The Davidson committee was critical that cross-subsidisation dominated Telecom's tariff arrangements: call charges cross-subsidised rental charges, trunk calls cross-subsidised local call revenue, metropolitan services cross-subsidised country services, and customers close to exchanges cross-subsidised customers remote from exchanges. In their final recommendations the Davidson committee questioned the principle of whether it ought to be the task of Telecom management to make arbitrary judgments about the priorities of cross-subsidisation, or whether it was a matter for the elected parliamentarians.

Davidson recommended that Telecom should shift from uniform pricing to cost-related pricing practices. In more recent years, competition policy has demanded there be significant moves towards 'rebalancing' Telstra's tariffs so that prices to consumers are more directly related to costs. This policy implies 'de-averaging', where prices are charged on the basis of the cost of providing individual services rather than being based on the average cost of a similar cluster of services. These pricing issues have been 'hot' topics for debate in most telecommunications policy fora of the world and are likely to remain difficult to resolve for a long time to come. In June 1989, pricing regulation was passed to Austel which was given greater authority over a carrier's pricing where the carrier was deemed to be 'dominant' in a particular market. In 1992 Austel declared Telstra to be dominant in four major markets, the local, STD, IDD and mobile markets. Subsequently, however, following the vigorous competition provided by Optus and Vodafone in the mobile market, Austel determined (April 1994) that Telstra was no longer dominant in the mobile market. In September 1994 Telstra challenged Austel by advising the regulator that it no longer considered itself dominant in the international services market, that it intended to charge prices below its tariffed rates for some of its IDD calls, and that consequently it should no longer be subject to regulation for its IDD pricing. Austel responded by conducting an inquiry into the matter which concluded that Telstra remained dominant in the international market and that its IDD prices were to remain subject to regulation. Telstra's response was to mount a challenge against Austel's ruling in court. Here, then, was the new telecommunications environment at work—an incumbent carrier standing its ground to maintain its market pre-eminence, a regulator committed to the implementation of competition principles to get consumer prices down, and the new entrants complaining about Telstra's alleged excessive market power.

Governments also use price capping regulation as a control mechanism on carriers, a technique that the British government first introduced in 1984 into telecommunications to curb British Telecom's post-privatisation pricing excesses. A price cap formula (CPI − X) per cent per annum involves two main variables, the Consumer Price Index (CPI), and an X factor determined by the government through the regulator to ensure that prices will fall in real terms. If, for instance, inflation was running at 8 per cent, as measured by the CPI, and the X factor was set at 3, the prices would be required to fall by 8 − 3 = 5 per cent, and the real inflation adjusted prices by X = 3 per cent per annum. This means that consumers benefit by real lower prices and that the carrier must improve its productivity by at least the value of X per annum to maintain their profit margins. The eras of different price caps are summarised in Table 4.4.

The post-1 July 1997 legislation deemed that:

- The price cap legislation would continue until at least 31 December 1998.
- The price cap regime was to be reviewed in 1997/98 before the expiry of the regime.
- The ACCC was required to report to the Minister annually on prices for specified carriage services.
- The ACCC was responsible for administering the price cap regime.

Table 4.4 Price regulation of Telstra services, 1989–98

Period and price cap	Basket of services	Other constraints	Notification and disallowance
Pre-competition 1989–92 CPI − 4%	Rentals Local calls STD calls IDD calls		Connections Public pay phones Directory assistance Mobile services Leased lines
Post-competition 1992–95 CPI − 5.5%	Rentals Local calls STD calls IDD calls Connections Mobile services Leased lines	• Basket of three services (connections, rentals, local calls): CPI − 2% • Basket of all STD calls: CPI − 5.5% • Basket of all IDD calls: CPI − 5.5% • Connections, rentals, local calls, individual STD calls: at CPI	
1996–98 CPI − 7.5%	Rentals Local calls STD calls IDD calls Connections Mobile services Leased lines	• Connections, rentals, individual STD and IDD calls: CPI − 1% • Local calls, public pay phone charges: no nominal increase 1.1.96 to 31.12.98	Directory assistance Interconnection for service providers

Sources: Austel, Australian Communication Authority.

Many governments during the past decade have used price capping regulation of the retail prices charged by a dominant carrier to achieve price reductions for consumers. Patrick Xavier has pointed out that 'price caps are professed by regulators to be a regulatory instrument to be applied temporarily in the "transitional stage" and that the aim should accordingly be to reduce the price cap coverage as competition in the service areas developed'.[11] However, he points out that his research on price capping in 27 OECD countries indicates that 'the price cap coverage over services has actually increased steadily . . . and the price cap formula has been steadily tightened with each review of the formula . . . the need for stringent price caps seems an admission that competition is not yet effective'.[12] The Productivity Commission's estimate of the 'welfare cost' of price capping was just over $400 million per annum.[13] Price capping may be good politics, it seems, but bad economics.

In Australia, in an era when the belief in competition has become so entrenched, Allan Brown raises the interesting proposition that for the duopoly period 'it has been regulation not competition that has been mainly responsible for Telstra's price reductions'.[14] He points out that the Bureau of Transport and Communications Economics (BTCE) has noted that 'falls in price achieved since the introduction of competition in 1992 have been similar to the declines experienced prior to the introduction of competition'.[15] For Brown, the effect of competition during the duopoly phase was more to influence the extent to which the regulated overall price reductions were distributed across the range of Telstra's products. He argues that price reductions tended to concentrate on mobile, STD and IDD calls, and to benefit primarily large-volume, high-expenditure customers. He concludes that 'competition has caused the reductions to be distributed unevenly across subscriber groups'.[16]

Pricing policy is seen as one of the most contradictory arms of government policy in the post-liberalisation era. The argument is that holding down prices by regulation can reduce the attractiveness to possible new market entrants and thereby reduce competition. However, strict pricing controls, initiated by government, remained in place until 31 December 1998, and the Minister for Communications, Richard Alston, later extended the price capping regime to 30 June 1999. In summary, the price controls:

- require Telstra to reduce the price of a basket of its main services (connections; line rentals; local, trunk and international calls; leased lines and mobile services) by 7.5 per cent in real terms in each calendar year in which a price cap regime operates;
- require Telstra to reduce its standard prices for residential customers, for connections, line rentals, trunk and international calls, by 1 per cent in real terms in each calendar year in which a price cap regime operates;
- prohibit the price for untimed local calls from rising above the current charges of 25 cents for calls made from a residential or business phone, and 40 cents for calls made from a public pay phone;
- require Telstra to obtain the prior consent of the ACCC where Telstra proposes to increase a charge that is subject to the price

control arrangements by more than the change in the CPI during a year; and
- make Telstra's charges for directory assistance subject to notification and disallowance by the Minister.

The Minister added that following consultation during the first six months of 1999 the government would make a decision on 'the structure and scope of the price cap scheme' that will apply for the 1999/2000 financial year and beyond.[17]

In the early period of the post-1 July 1997 open competition era, there was considerable anecdotal evidence that competition had begun to bite and that prices were beginning to fall further for select services. In opening the May 1998 Australian Telecommunications Users Group (ATUG) conference, Warwick Smith, who spoke for the government in the Lower House on communications, said:

> [M]ore than a dozen carriers have so far been licensed but the most tangible consumer benefit has been price reductions. The very day the new regime began, 1 July 1997, AAPT announced that it would offer up to 60 per cent off the current cost of national and international long-distance calls. Other players were hot on AAPT's heels. A number of service providers—including Global One, Primus and WorldxChange—offered calls to the United Kingdom, the United States and New Zealand for as little as 37 cents per minute in peak weekday periods. Innovative offerings included, for example, Northgate Communications in Ballarat and the major capital cities, who offered untimed calls to Los Angeles, Hong Kong, London and Auckland for $4.95. Both Telstra and AAPT have offered a capped $3 call around Australia on weeknight evenings. Optus has recently announced that it will match this offer.[18]

(Northgate subsequently went insolvent.) Analysis conducted by the Department of Communications, Information Technology and the Arts into price levels in the more competitive areas of the Australian telecommunications markets showed that prices had been used as a competitive weapon in the immediate post-deregulation period. The conclusion was that 'the range of charges now offered for particular calls has widened, with some operators offering services at rates less than half those of incumbent carriers and Optus no longer "shadowing" Telstra's rates. The incumbent carriers remain the more expensive operators in the marketplace.'[19] Telstra has long been accused of maintaining expensive local call rates by international standards. However, the work undertaken by Analysys Ltd (for Telstra), as reported by John de Ridder, questions the accuracy of this widespread assertion. Adjustments need to be made in looking at local call pricing because there are such wide variations in how local calls are defined between countries, including countries such as Canada which offer 'free' local calls. Hence, credible analysis must 'look at local calls and access in turn and consolidate them into a normal comparison'. The conclusion was that Telstra was second only to Singapore in cheapness of local service to consumers.[20]

Competition, it seems, has begun to replace regulation in driving down consumer prices in Australian telecommunications. The early days of open competition are certainly seeing some flowers blooming.

Universal service obligations (USOs)

The historical justification for the public ownership of telecommunications centred on the assumption that community service obligations, especially the provision of telephone services to remote Australia on a non-discriminatory and uniform basis, pay phones, access to directory assistance and emergency numbers, were best provided by the publicly owned common carrier through internal cross-subsidies. Hence the losses incurred in providing uneconomic 'rural and remote' services were traditionally cross-subsidised by Telecom's long-distance, international and inner metropolitan services. An economist is quick to point out that phone prices to consumers in non-rural areas had to be inflated to pay for rural services, and that community service obligations (CSOs) exclusively carried by a common carrier inhibited private sector investment and competition.

There can be little doubt that up until the 1990s, the pre-competition era, the policy objective of providing universal services at affordable prices to most Australians was successfully realised. The Postmaster-General's (later Telecom) Community Telephone Plan in 1960 had admirable clarity of purpose—to provide everyone in Australia with access to the national telephony network. In 1939 there were less than 500 000 telephone services in Australia, growing to two million by 1960. As a result of the Community Telephone Plan, 62 per cent of households were connected by 1975, and by 1996 an impressive 96 per cent of Australian residential households had a telephone.[21] The Community Telephone Plan of 1960 also set out to extend STD nationwide to allow subscribers to dial their own calls to any other subscriber in Australia; the STD household access rate improved to 99 per cent by 1981.

As we became more dependent on telecommunications products and services, a series of tough social justice policy questions arose, neatly summarised by Holly Raiche:

- How can we define in precise terms the nature and quality of the service that should be available?
- Should the service reach all parts of the country, and if so how?
- Is the service affordable, particularly for those on low incomes?
- Should government policies cater for people with special needs, such as people with disabilities?[22]

With the introduction of competition during the 1990s, these questions increasingly focused on who was responsible for telecommunications social policy, and who paid. The 1989 legislation, in pre-duopoly days, introduced the concept of community service obligations, which required Telecom to ensure:

a. that in view of the social importance of the standard telephone service, the service is reasonably accessible to all people in Australia on an equitable basis, wherever they reside or carry on business; and
b. that the performance standards for the standard service reasonably meet the social, industrial and commercial needs of the Australian community.

Telecom was to both provide and fund CSOs through its cross-subsidies.

However, the policy of the public common carrier being charged with the exclusive responsibility for CSOs was obviously inappropriate with the introduction of competition, because Telecom would have unfairly borne the exclusive responsibility for non-economic services. Telecom found itself under great pressure to 'open its books' to wider scrutiny about the true nature of its complete range of cost subsidies, and questions were asked about whether they were a convenient way of covering up alleged inefficient financial practices within the organisation. The pressure was on to define precisely what non-economic services could be justified by cross-subsidies, symbolised by the shift in rhetoric away from the use of the broad term 'community service obligations' to the narrower term 'universal service obligations' (USOs).

In framing the legislation, in both the 1991 and the 1997 Telecommunications Acts, the policy-makers were aware that the areas of Australia still needing network capacity for telephones, especially those neglected outlying regions, were hardly going to get high priority from any of the new carriers, or from anyone else. So provisions in the legislation, called universal service obligations, were enacted to ensure that a designated carrier would be required 'to provide a given level of service to an area irrespective of the fact that the service is commercially unprofitable' (section 288(1) of the 1991 Act). Here the USO is defined as the obligation:

a. to ensure that the standard telephone service is reasonably accessible to all people in Australia on an equitable basis, wherever they reside or carry on business;
b. to supply the standard telephone service to people in Australia;
c. to ensure that pay phones are reasonably accessible to all people in Australia on an equitable basis, wherever they reside or carry on business; and
d. to supply, install and maintain pay phones in Australia.

Telstra was declared as 'the USO carrier' but was able to seek compensation via a universal service levy, whereby Optus was obliged to reimburse Telecom a portion of the total USO cost calculated according to its share of timed traffic across the network. Austel data show that in each of the three years 1992–95, Telstra absorbed over 94 per cent of the calculated total net universal service cost, with Optus paying only a small amount of the non-Telstra portion.[23] The *Telecommunications Act 1997* left the levy in place, and extended it to apply to other new carriers, but as a result of some effective lobbying from Telstra, introduced the prospect of a tendering process intended to reduce Telstra's USO prime responsibility.

Universal service obligations policy, to date, has essentially been restricted to access to telephony. Yet if we are living through an information revolution, why should access to the service be restricted to telephony only? The leading telecommunications consumer lobby group, Consumers' Telecommunications Network (CTN), argued that the government only addressed geographical location in its USO considerations and called for 'a comprehensive and future oriented definition of USO', the components of which ought to be:

a. universal geographical availability;
b. universal accessibility;
c. universal affordability;
d. universal technical standards; and
e. universal telecommunications and participation in society.[24]

We need to rethink USOs not as universal service *obligations* but as universal service *opportunities*.[25] Given the growing significance of these new information services to the way so many Australians run their lives, why can't we revise our USO thinking to explore facsimile, and data services, especially the Internet? How, too, might we begin to think in terms of better access to the non-fixed network services, such as mobile telephones? There are some extraordinarily promising opportunities that could emerge from broader thinking than merely a legislated definition of USO as telephony.

Telstra: Sale of the century

Telstra is the jewel in the crown of Australia's public assets and its network is the nation's most valuable strategic infrastructure. The possible privatisation of Telstra has been a hotly contested political issue during the past decade or so, with Australian political parties keenly aware of the conflicting passionate positions held by the public and various vested interests, and of the serious electoral risks for any major political party moving to privatise the company.

Telstra is a remarkable telecommunications company in many ways. It is ranked at number 16 in the world on its revenue performance, well up among the world's top telecom giants.[26] Just over a quarter of a century ago (1974), Telecom was a mere $853 million revenue company, yet by 1991 (as AOTC) it had become a $12 billion company, with an after-tax profit for the first time of over $1 billion. In just over a decade and a half this Australian telecommunications corporation had grown—under public ownership—from being less than a billion dollar *revenue* company to become a billion dollar *profit* corporation. By June 1998 Telstra had grown to a $17.3 billion company with net after-tax profit in excess of $3 billion, having declared total dividends of $1.8 billion, fully franked.[27] Telstra is the biggest corporate investor in the Australian economy: it paid the Commonwealth and state governments a staggering $6027 million in dividends, taxes and charges in 1996/97 and employed 66 109 full-time staff as at 30 June 1997 (compared with 76 522 in June 1996).[28] Telstra, the communications company Australians love to hate, is a great corporate success story.

Privatisation is essentially the transfer of supply of a good or service from a government (public) agency to a non-government (private) body. A spate of privatisation of telecommunications companies has occurred around the world since the mid-1980s in several different ways:

- All of the company can be sold, such as Chile's ENTEL in 1988.
- The company can be progressively sold, such as British Telecom shares which were sold in three tranches, in 1984, 1991 and 1993, for a total net sale value of US$22.8 billion. Although the UK government has no direct ownership of British Telecom, it can still exert influence through policy, licensing and regulation.
- Sale of the company, but the government retains a 'golden share'

or veto in matters of public interest, such as the New Zealand government's 100 per cent private sale of Telecom New Zealand to Bell Atlantic and Ameritech for US$2.5 billion in 1990.[29]

Australia has been 'slower' to privatise its common carrier than most of its major trading partners. It was always going to be politically difficult for any of the Australian Labor governments, in office from 1983 to 1996, to move towards the privatisation of Telstra. Public ownership is embedded in the Labor Party's ideological roots with its strong trade union base of support. Soon after the re-election of the Labor administration in July 1987, Prime Minister Bob Hawke floated the prospect of possible sell-offs—an action initially widely regarded as ideological treachery within the Labor movement. However, at the national conference in September 1990 Labor took the plunge and agreed on a modest privatisation program which included the Commonwealth Bank, Trans Australian Airlines and the international airline carrier, Qantas, as well as Aussat. What followed was a spate of privatisation ventures (see Table 4.5).

Although many senior Labor Party figures believed it was inevitable that Telstra would be sold, Labor never risked its electoral support on a platform of Telstra privatisation. Labor did, of course, sell Aussat, the domestic satellite company, to Optus to create the carrier duopoly in 1991, although part of this strategy was to lock OTC into Telecom to prevent its separate privatisation.

John Hewson, Liberal leader during the 1993 election campaign, was the first Australian political leader to test the electorate on telecommunications privatisation, but he was defeated in a campaign dominated by the goods and services taxation issue. Three years later

Table 4.5 Australian industry privatisations in the 1990s (pre-Telstra)

Public asset sales	Value (A$bn)
1991 Commonwealth Bank (30%)	1.3
1993 Commonwealth Bank (19.9%)	1.7
1994 Commonwealth Serum Laboratories	0.306
1995 Qantas	1.46
1996 Commonwealth Bank (50.1%)	2.4

Sources: 'Selling off the silver', *Age*, 16 March 1998, Business, p. 3; and 'Telstra privatisation: Business hamstrung by politicians', *Age*, 13 November 1998, Business, p. 3.

John Howard, Liberal leader, won office comfortably at the 1996 election in which his policy platform included the one-third privatisation of Telstra. There was initially great difficulty in getting this through a hostile Senate, but eventually two independent senators agreed to pass the legislation in December 1996. In this sale of the first one-third of Telstra, the government earmarked $1 billion of sale proceeds for major national environmental projects, which later emerged as projects targeted to their electoral advantage. Under the Howard Coalition government the sale of one-third of the company was formally launched in September 1997 by the Finance Minister, John Fahey, who declared that small investors would receive strong preferential treatment. Telstra's float was listed on the stock exchange on 17 November 1997. Extraordinary trading followed, and small investors who had paid $3.30 at the beginning found that their shares topped $9.20 in February 1999. By 1998 Telstra was capitalised at $104 billion, doubling its paper value during 1998, easily the best-performing stock in Australia.

In March 1998 the one-third sale of Telstra had been such a brilliant stock market success that Howard proposed on behalf of the Coalition to sell the other two-thirds of Telstra if re-elected. Labor in Opposition had declared a policy of no further sale of Telstra. The government actually presented Parliament with legislation to implement the sale of the remaining two-thirds of Telstra before the election, but it was lost on the casting vote of an independent senator in July 1998. Mr Howard also faced dissent within his own ranks about a Coalition policy of full Telstra privatisation. Under pressure the revised Coalition policy was to hold the level of privatisation at 49 per cent pending service performance reviews to appease rural and other party interests. Labor in Opposition again declared a policy of no further sale of Telstra.

Following the re-election of the Coalition with a reduced margin in October 1998, the government introduced further privatisation legislation in November 1998, with the first section of the bill to facilitate the sale of a further 16.6 per cent of Telstra, but with legislative provision to eventually sell down the rest. The sell-down was conditional on the outcome of an independent review of Telstra's service standards, particularly in rural areas. In June 1999 the Senate approved the sale of a further 16.6 per cent of Telstra with the support of two independent senators, Brian Harradine and Mal Colston, leaving 50.1 per cent in government ownership. The deal to secure these votes included a $314 million Telstra 'social bonus' package of which $100 million was earmarked for Tasmania, Senator Harradine's home state, in addition to a previous $40 million grant to develop that state as the 'Intelligent Island', and a further $15 million computer connect program for Tasmanian schools. Opposition Leader Kim Beazley attacked the deal as 'a shameless bribe', adding that it was 'one of the worst cases of pork barrelling we've ever seen in Australian politics'.[30] The *Australian Financial Review* disagreed and editorialised, '[T]he macro-economic benefits . . . are much more important than the pay-offs to the independent senators who held the fate of Telstra in their hands.'

What is the case in principle for the privatisation of Telstra, and how valid or credible are the respective arguments for and against a change of ownership? The key arguments are canvassed briefly below.

Efficiency

The Howard government's prime justification for the sale of Telstra was that it would 'create a new efficient Telstra'. According to Richard Alston, as the Minister for Communications in 1998, Australia was the only telecommunications carrier that was not privatised among the top twenty telecommunications companies, by revenue, in the world. While telecommunications has gathered great pace over the past decade or so, it is not actually the case that all countries had committed to full privatisation. In fact, of the 29 OECD countries in mid-1998, only ten had fully privatised their dominant telecommunications carrier, or have policy commitments to do so.[31]

Although privatisation is widely advocated as an absolutely necessary component of reform of public utilities, the empirical case linking the change to private sector ownership to greater efficiency has not been proved. Analysis of the relative operating efficiency of public and private organisations is inconclusive. Allan Brown's critique of empirical evidence on the relative efficiency of many public and private enterprises raises the methodological problem of on what basis efficiency is assessed. He is critical of the measurement of productivity based on sales per employee and per asset, as opposed to using total factor productivity as the basis of measurement.[32] There is extensive research and literature on the issue of the relative efficiency of public and private sector organisations, but it has not been shown that one form of ownership clearly leads to greater efficiency than the other.

An Australian Senate Committee (with a minority of pro-privatisation government representatives) spent months examining this issue and concluded that 'the strongest advocates of privatisation have been the Telstra management and market analysts, two groups most likely to directly benefit from privatisation . . . [the committee] rejects the assertion that there is a direct and inevitable link between economic performance and type of ownership'.[33]

Privatisation is, of course, a key ingredient of economic rationalism and its rejection of government involvement in industry practices in favour of a belief in market-based decision-making. Simon Domberger neatly summarises these arguments: '[A]dvocates of privatisation argue that only the incentives generated by capital market pressures . . . will ensure that assets are profitably employed . . . [and] emphasise that private ownership and its concomitant goal of profitably lead to a continual search for efficiency in production and responsiveness to consumer demand.'[34] On the other hand, public enterprises have not historically given productive efficiency a high priority, and have had non-commercial activities, such as community service obligations, prominent in their charter of responsibilities. Moreover, as Dianne Northfield has pointed out, '[P]rivate organisations' achievement is based on a system of incentives including management/staff bonuses, shares and stock options, accountability to board members and shareholders concerned with capital efficiency, and subjection to capital market scrutiny and discipline, and a range of disincentives to poor performance including the threats of takeover and bankruptcy.'[35] The value of these forms of managerial accountability, however, was brought into question by the notable excesses of corporate behaviour during the late 1980s.

Telstra's alleged inefficiency is a favourite dinner party debating subject for many Australians who can unearth loads of anecdotal evidence about alleged sloppy practices, especially on billing and maintenance work. OECD research on comparative call tariffs indicates that we are in the middle of the ruck of common carriers for charging, and BTCE research argues that Telstra's productivity performance lags behind international best practice.[36] However, since 1991, when the corporation was subjected to Government Business Enterprises' (GBE) 'competitive neutrality', Telstra's revenue earnings, after-tax profit, dividend payments to government, and return on assets, investment and equity have been impressive. Since corporatisation in the 1980s, and the introduction of greater competition in the early 1990s, Telstra has undoubtedly achieved considerable productivity growth—under public ownership.

And what of the relationship between privatisation and deregulation? *Privatisation*, the sale of a public asset to private interests, is related to, but different from, *deregulation*, the freeing up of the operational rules which govern the industry. The term 'deregulation' best applies to a circumstance where a monopolist carrier faces competition for the first time. Often, deregulation, in practice, is merely a new regulatory regime to replace the old, and could be more accurately described as *reregulation*. It is possible to deregulate a telecommunications industry, as well as introduce new players through competition, and yet not privatise the common carrier. This was essentially what happened during the period of the five Hawke–Keating Labor governments (1983–96) with the introduction of new common and mobile carriers, such as Optus and Vodafone respectively, and freer market rules through deregulation, including interconnection, permitting resale and allowing freerer network attachment, but Labor did not privatise the common carrier, Telstra. When a market is highly deregulated, highly competitive and privatised, the term 'liberalisation' is applied.

The economic literature tends to be more convincing on the benefits of competition and deregulation than it does on the benefits of privatisation. Simon Domberger and John Piggot argue that 'asset sale may enhance the beneficial impact of deregulation but asset sale in the absence of deregulation is unlikely to improve efficiency, and may introduce additional market distortions'.[37] Similarly, Vickers and Yarrow concluded from their research that 'the effects of privatisation in any particular context will, therefore, be highly dependent upon the wider market, regulatory and institutional environments in which it is implemented'.[38]

In short, then, the benefits of competition and deregulation appear to outweigh the alleged benefits of privatisation.

Financial purpose

Another conflicting objective of privatisation is to determine the purpose of the substantial capital injection of funds paid to the government. Is the refinancing of Telstra intended to benefit the company and the industry, or is the payment primarily to the benefit of the country through a one-off debt retirement? In an age of necessary fiscal restraint, with limits on public sector borrowings by Commonwealth and state governments, an argument put in favour of privatisation is that the necessary funds for Telstra to finance new infrastructure, develop new services and introduce expensive technological innovation can only be attained through private sector recapitalisation at a level which will make Telstra internationally competitive. However, the privatisation debate has centred more on the politics of debt retirement by the government than on the alleged private capital stimulation that could become available to a fully privatised Telstra.

The one-third float of Telstra in November 1997 raised an impressive $14 billion for the government, and attracted investment from

1.9 million Australians, 560 000 of whom entered the share market for the first time, taking share ownership in Australia to an estimated 40 per cent of the population. By mid-February 1998, Telstra's shares, then priced at $3.56 per instalment, valued Telstra at $70 billion all up, which put it in the world's top ten telecommunications companies, although it serves only 0.3 per cent of the world's population. Buoyed by this remarkable success the Prime Minister, John Howard, subsequently invoked the ghost of Bob Menzies when he announced in March 1998 the government's intention to float the other two-thirds of Telstra if re-elected, because, he said, '[I]t is my goal [to] make Australia the greatest share owning democracy in the world.'[39] The Federal Minister for Finance, John Fahey, told Parliament that the proposed sale of the rest of Telstra could reduce Commonwealth debt by the sale proceeds of up to $40 billion, which could see Commonwealth debt reduced to just $25 billion, down from the $100 billion of public debt the Liberals inherited from Labor on gaining office.[40] Later, the Treasurer, Peter Costello, in handing down the 1999 Budget, said that Australia could enter the next century with no Commonwealth government debt if the Parliament would agree to the full privatisation of Telstra.

In March 1999 the out-going Managing Director of Telstra, Frank Blount, strongly advocated the full privatisation of Telstra on these financial grounds:

> Full privatisation will lead to increased tax payments by Telstra to the government, economic growth flowing from the company's success, inflows of foreign exchange, the near elimination of Commonwealth debt, and other benefits that far outweigh the dividends Telstra would pay to future governments which retained ownership.[41]

Serious doubts, however, have been raised about the validity of these financial arguments in terms of the long-term national interest. At the time of the first float, John Quiggin challenged what he regarded as the government's opportunistic financial arguments, on the grounds that it was short-term thinking for the government to focus on short-term gains in interest payments at the expense of long-term dividends. Quiggin argued that, for the remaining two-thirds sale of Telstra,

> assuming the sale price of $45 billion, the government could reduce its interest payments by about $2.5 billion per year in 1998 (and every year in the future) by selling Telstra and using the proceeds to repay debt. The government would, however, lose its continuing present dividend claim from two-thirds of Telstra's earnings. In 1998 the value of this claim was $2 billion, so the government was ahead by $500 million. But Telstra's

profits have been growing rapidly and the market obviously expects that this growth will continue, whether or not Telstra is fully privatised. A conservative assumption is that Telstra's profits will grow in line with nominal GDP, that is by around 5% per year. On this assumption, the short term net benefit to the government would disappear within five years, to be replaced by a steady stream of losses . . . The sale of the remaining two-thirds of Telstra would be a mistake.[42]

This prime argument, that the money saved in short-term interest payments resulting from the sale would be much less than the long-term dividends paid to government under public ownership, was later forcefully argued by columnist Kenneth Davidson of the *Age*:

Since Telstra has been privatised the net profits attributable to the one-third private shareholding have totalled $1.6 billion over 18 months. Over the same period, if the government had not sold the shares, and, as a consequence, retained $8.3 billion in debt, the additional interest payments on government debt would have amounted to $750 million. Thus, in the first 18 months, government revenue is $850 million worse off as a result of the sale . . . The bulk of Australians have been ripped off once in order to pump a $750 million a year bonus into the new private owners of Telstra shares.[43]

There are so many financially unpredictable variables in these debates. Not the least of these is the assumption that the future level of taxation paid under private ownership would be comparable to the tax and dividends paid under public ownership. Where is the evidence to support this in Australia? In recent years, times of considerable fiscal restraint, Australian governments have become highly dependent on Telstra revenues to fulfil their annual budgetary responsibilities. Long term, there may be serious social national consequences if there were substantial reductions in Telstra revenues paid to government. Privatisation may be smart politics, and popular among its select shareholders, but not necessarily good long-term policy in the national interest.

National strategic purposes

The full privatisation of Telstra would hand the power and prime authority of the company over to select private interests. This raises many issues and concerns about accountability and changing priorities under a management driven by intense profit maximisation. Would a fully privatised Telstra dominate the telecommunications market? In March 1999, major companies who had entered the market since it was opened up to further competition warned that further privatisation of Telstra would reduce the government's capacity to curb

Telstra's anti-competitive behaviour. In a letter to Senator Richard Alston, ten telecommunications companies—Cable and Wireless Optus, AAPT, Global One, Macquarie Corporate, One.Tel, PowerTel, Primus, MCI WorldCom, Worldxchange and BT Australia—argued that 'present government ownership of Telstra provides some brake (albeit small) on the company's excessive and anti-competitive behaviour'.[44] What checks would be kept on intense profit maximisation in the pricing of services to consumers? This was a serious problem the Thatcher government faced soon after it fully privatised British Telecom. How will infrastructure and services be developed for all of Australia, including rural Australia, under full private ownership? The proponents of privatisation argue that an appropriate regulatory regime can ensure that vexed matters such as rural services and fair pricing are adequately addressed. There are, however, grave fears held among rural consumers, some of whom have lost their local bank, local school, local post office and local rail station, that telecommunications services will go the same way in this new drive for urban global efficiency. No government can really assure these rural interests that the regulatory regime set in place at the time of privatisation with good intentions will remain in perpetuity. The new owners will be able to pressure a government in many ways for changes in regulation once they are firmly in harness.

And what of local industry? Australia has a significant local electronics manufacturing industry built around its world-class telecommunications infrastructure. Local industry development is most unlikely to be a priority for a fully privatised carrier, especially if it eventually has a high percentage of foreign ownership. Regulatory policy guidelines may have licence terms and conditions imposed on the carrier about local industry development, but these will be much more difficult to see achieved under private ownership than under public ownership. Moreover, a privatised new Australian carrier already competing in export markets may have a conflict of interest where it has pre-existing relationships in the same market in which there are development prospects for Australia.

Privatisation fundamentally changes the locus of power in Australian telecommunications.

Australian telecommunications: Reviewing the new balance sheet

Australian telecommunications policy changed more in the last decade of the twentieth century than during any other period of its history. Essentially, economic policy has come to subsume and subvert social policy. For decades the driving force of Australian telecommunications policy was an overriding social objective, to deliver universal service at affordable prices. Australia achieved that goal faster than almost any other nation in providing high levels of telephony access to consumers through building and managing one of the best publicly owned and controlled telecommunications networks in the world. Inevitably, Australia had to shift towards finding its own best solution in a world where our trading competitors were working towards a more privatised, competitive, deregulated, liberalised telecommunications environment. Government policy shifted during the 1990s away from public carrier monopolies, with Telecom as dominant domestic carrier, OTC as international carrier and Aussat as domestic satellite carrier, to the 1991 model of duopoly based on Telstra and Optus, to a one-third privatised Telstra in an environment of open competition post-1 July 1997. All of this was achieved with commendable legislative efficiency given the degree of change. But what were the fundamental objectives behind these changes, and what do we know about the outcomes?

The new Australian telecommunications environment has seen the emergence of a plethora of new players, collectively sharing about 20 per cent of the overall telecommunications market. The incumbent carrier, which for so long was accused of being sloppy and inefficient, is trading at record levels of profitability, and offers the government, and select Australians as new shareholders, rich pickings for full privatisation. Some of the new entrants have made their mark, but none of them has yet been able to stamp their clear authority on this vibrant, aggressive service marketplace. Consumers have had more choices of operators and services, lower telephony costs for interstate and international phone calls, and more responsive suppliers.

The institutional players find themselves ill at ease with the new model with its inherent tensions of conflicting policy objectives and the insecurities of a highly competitive marketplace. Telstra management has to grapple with what it sees as a contradiction between the government's role as regulator for the telecommunications industry as a whole, together with its role as owner of the nation's most profitable company which it wants to privatise. Apparently these tensions could be resolved by full privatisation. Meanwhile, the management of the new carriers, notably Cable and Wireless Optus and AAPT, vigorously protest that only about 20 per cent of the marketplace, in long-distance and international phone calls, has been freed up in Australian telecommunications.[45] The Service Providers Action Network (SPAN) has regularly expressed its disgust at what they see as the slowness of Telstra to open up its data networks to more competition. Ten new carriers, including AAPT, British Telecom Asia Pacific, Global One, MCI WorldCom, Macquarie Corporate and Primus, argued that the costs of interconnection for Telstra, which determine what its competitors pay it in interconnection charges—so vital in terms of creating effective competition—were about one-quarter of that claimed by the industry leader.[46] Since open competition was introduced, we have witnessed disputes over billing which have cost millions of dollars, and a $900 million damages claim filed by Optus against Telstra in the Federal Court.

Three major trends seem likely to chart the future direction of Australian telecommunications:

1. With its plethora of new players, the Australian telecommunications marketplace has become more aggressive and much more legalistic, with much energy focused on Telstra's alleged dominance. The arbitrator of brawls, the Australian Competition and Consumer Commission, has emerged as the de facto policy architect for the future of Australian telecommunications, rather than the elected national government.
2. The multiplicity of new players, carriers and service providers, initially attracted by the opening up of the marketplace, will probably shake down to fewer participants than now, although many more than ever before. The marketplace will become even more intensely competitive with cut-throat competition and low margins for all players.
3. Universal service obligations by carriers will come under greater pressure, local industry development as a policy objective is likely to be abandoned, and there are likely to be few long-term strategic infrastructure investments, especially by carriers, who will be forced to opt for short-term returns in a much less stable industry than ever before, despite its spectacular overall growth.

We have come to accept competition policy as the driving force for setting new directions, rather than integrated economic and social objectives defined by public policy. In one of the grandest ironies of deregulation, it is noticeable that Telstra's competitors, and the Australian Telecommunications Users Group, have made constant calls on the Minister to be more aggressive in policing the industry and to strengthen the powers of the ACCC. Clearly, Telstra is in for a tough time in future from the ACCC, and the carrier, in response, will strengthen its legislative apparatus. Apparently this shift to a highly legalistic orientation is a stage we must pass through before we reach a mature phase of competition.

The future collective success of participants in the Australian telecommunications marketplace will be vital to sustain Australia's national prosperity. The management view at the top in Telstra, supported by investors, urges that the serious tensions that have emerged between *competition policy* and *social policy* be resolved, and they propose the policy remedy—the full privatisation of Telstra. Economic/social tensions are inevitable, indeed a healthy indicator of good democratic practice, and will not be solved merely through change created by an economic instrument. Meanwhile, Telstra's competitors seem to see a different difficulty—the unresolved problem of Telstra's alleged dominance in the deregulation era. And what of those who need and use telecommunications—the Australian people? Allan Horsley has called for the communications industry to concentrate on genuine consumer needs and move on from its intense preoccupation with the supply side of the market. Horsley suggests, 'We have the technology. We have the infrastructure. We have more fibre optic cable than we know what to do with. But we have lost the plot on the end user.'[47]

Finding the plot for Australians, not only as consumers but as citizens, is the real challenge of the new telecommunications age.

5 Electronic nomads: Internet as paradigm

In fact, life in cyberspace seems to be shaping up exactly like Thomas Jefferson would have wanted: founded on the primacy of individual liberty and a commitment to pluralism, diversity and community.

Mitchell Kapor, 'Where is the digital highway really heading?', *Wired*, July/August 1993, p. 53

While the Internet beckons brightly, seductively flashing an icon of knowledge-as-power, this nonplace lures us to surrender our time on earth ... the medium is being oversold.

Clifford Stoll, *Silicon Snake Oil: Second Thoughts on the Information Highway*, Pan Books, London, 1996, p. 4

The concepts of 'paradigm' and 'paradigm shift' are derived from Thomas Kuhn's classic book of 1962, *The Structure of Scientific Revolutions*, in which he explored how the state of knowledge undergoes rapid periodic advances within and across scientific disciplines. Paradigms, according to Kuhn, are 'universally recognised scientific achievements that for a time provide model problems and solutions to a community of practitioners'.[1] For Kuhn, paradigms share two essential characteristics: their achievement is sufficiently unprecedented to attract an enduring group of adherents away from competing theoretical models of scientific activity, yet simultaneously sufficiently open-ended to leave all sorts of problems for those adherent practitioners to resolve. The Internet marries the elements of human mediated communication—speech, writing, vision, data and sound—but uses computers and telecommunications to unite them on one platform. The Internet's extraordinary growth and global reach of the platform in recent years, the passion of its adherents and its maze of unresolved issues all qualify it as a paradigm shift. The jury remains out, however, on whether it will fundamentally change the way humans communicate with each other.

The Internet is not a single physical tangible entity, but a global aggregation of networks which interconnect smaller groups of computer networks. It constitutes a vast array of computer information and services that can communicate with each other as if they were one giant world computing machine. There are many different methods of information exchange over the network. Although constantly evolving and difficult to categorise, the most common methods of communication on the Internet may be usefully grouped into six categories:

1. one-to-one messaging (such as e-mail);
2. one-to-many messaging (such as 'listserv');
3. distributed message databases (such as USENET news groups);
4. real-time communication (such as 'Internet Relay Chat');
5. real-time remote computer utilisation (such as 'telnet'); and
6. remote information retrieval (such as 'ftp', 'gopher' and the 'World Wide Web').[2]

The most prominent platform that runs on this network of networks is the World Wide Web (www), which offers text files, images, sound, animation and graphic inputs from an enormous variety of sources all over the world. The Web utilises a 'hypertext' formatting language, called HyperText Markup Language (HTML) to create documents, and a Universal Resource Locator (URL) to find them. By clicking a hypertext link the Internet user in Australia can locate and view information stored on computers all around the world. The term 'intranet' refers to a 'closed' manifestation of Internet technology: an internal network intended to be accessed only by the users belonging to that organisation. A huge company such as BHP, which needs regularly to contact its thousands of staff members scattered throughout dozens of countries, finds intranet a valuable internal management communication tool. The Internet, however, can offer 'global' e-mail and a vast electronic encyclopedia of up-to-date information accessible through search engines to anyone in the world who has access to the system.

The Internet is a myriad of networked interconnecting computers forming an essentially open communications medium. In the 1980s this was anathema to a computer industry where so many of the major corporations long attempted to become the de facto proprietary computer system. No single entity administers or controls the Internet. John Hindle, Vice President of Nortel, North America, has argued, 'At its core, the Internet is an open communications medium—open to any computing device, open to any communications media, open to any public or private purpose.'³ In 1996, when the US Federal Court struck down the *Communications Decency Act*, an attempt in part to control certain controversial content matters on the Internet in the United States, the court declared the Internet as the most participatory form of mass speech yet developed. The Chief Judge ruled that 'just as the strength of the Internet is its chaos, so the strength of our liberty depends upon the chaos and cacophony of the unfettered speech the First Amendment protects'.⁴ The diversity of networks, and their decentralised and distributed ownership, has facilitated the pluralism of expression that is the hallmark of the Internet culture.

Survey data about who uses the Internet varies widely, and depends upon how the term 'Internet user' is defined. Users may be someone who has accessed the Net only once, but may not have actually used it, as well as someone who spends half of their day Net 'surfing'. Also, people use the Net for different purposes, and one site may be used by a variety of people. Early research into user numbers and profiles conducted in the United States in 1995 found that 67 per cent of those with Internet access were male, over half of them aged 18–34, with a median household income of between US$50 000 and US$75 000, with most predominant employment categories being engineering, sales and education.⁵ A different survey, again for the United States, in 1995 found that 65 per cent of users were male and affluent (average household income US$62 000), although older than indicated by the former survey (average age 36).⁶ So the evidence from the United States, the world's biggest Internet user, is that the Internet emerged as a medium of communication for predominantly the most educated and affluent male segment of the population who live in the most sophisticated metropolitan areas.

Internet access in Australia has grown rapidly since the mid-1990s. The www.consult market research group estimated the number of on-line users in Australia in March 1997 to be about 800 000, that most users of the World Wide Web had only begun to access it in the first half of 1995, but that the majority accessed it at least once a day. The typical Australian Internet user was then profiled as a tertiary qualified male, aged 20–44, with an average income twice as high as the national average.⁷ By August 1998, according to the Australian Bureau of Statistics, some 1 245 000 households, or 18 per cent of all Australian households, had access to the Internet from home, a staggering increase of 48 per cent from February 1998.⁸

An estimated 4.2 million Australian adults used the Internet between August 1997 and August 1998, representing over 23 per cent of the total population, with 1.9 million users accessing the Net from home and a further 1.9 million from work. The same research suggested that a total of 425 000 adults used the Internet for purchasing/ordering goods and services in the twelve months to August 1998 and that the average number of transactions was three. Of the four million adults who had accessed the Internet in the twelve-month period, nearly 350 000 had made an on-line purchase. The study found that Australians were reluctant to pay their bills or transfer funds over the Internet, with just 1 per cent choosing to do so between May and August 1998. By November 1998 nearly 19 per cent of Australian households had home Internet access—four times the number than in 1996![9] It seems that the community acceptance of on-line commercialisation, especially consumer financial transactions, has a long way to go.

Origins and growth of the Internet

In the post-Second World War era, several new communications technologies emerged from major investments in United States' defence programs—namely computers, communications satellites and the Internet. Paradoxically, America's militia have generated some of the world's most useful and constructive technologies for peaceful purposes! The Internet has passed through three major phases of development.

In *Phase One*, from the late 1960s to the early 1980s, the closed US defence development phase, the US Defense Department's Advanced Research Projects Agency (DARPA) set out to design a communications system that could survive a nuclear attack on the United States. Packet switching technology made it possible to design highly decentralised communication networks whereby if any communication point went down during transmission the information could be rerouted to other computers, which in turn could be linked to other computers. A communications system emerged with no centre; rather, a decentralised, self-maintaining series of links between computer networks, with the automatic ability to reroute communications elsewhere if one or more links were destroyed. At this stage the computers connected to the Net were situated in US military bases or research centres, and only available to a defence elite who used the system for e-mail transmissions. The early Net was used only by computer experts who had to learn a complex system. The first such network was named Advanced Research Program Agency Network (ARPANET), and was initially only available to research centres associated with the US Defense Department. Increasingly, though, scientists came to use it widely and the system and its funding was taken over by the US National Science Foundation. DARPA, the US defence agency which had designed this information network that was inherently difficult to control, divested the original network and it moved away from its original military strategic purpose. American universities, initially in the south-western states, took over the development of the Net under a contract let by the renamed Advanced Research Projects Agency (ARPA).

Phase Two, from the early 1980s to the early 1990s, saw academic and public and private sector research bodies playing leading roles in the extraordinary network and software development of the Internet. Over its short history the Internet has seen some significant collaboration among cooperating parties. As the commands for e-mail, FTP and telnet became standardised, it became easier for non-technical people to access the nets. This was an era of remarkable technological innovation: at the Massachusetts Institute of Technology's Lincoln Laboratory, Stanford Research Institute, Palo Alto Research Corporation (funded by Xerox), AT&T's Bell Laboratories and at Rand Corporation. *Gopher*, which enables the user to click on a number to select a particular menu, was produced at the University of Minnesota, and *Mosaic* at the University of Illinois. Three students at Duke University in North Carolina created a modified version of the Unix protocol that made it possible to link up computers over a telephone line. In 1989 a significant breakthrough took place which made the nets easier to use. Tim Berners Lee and others at the European Laboratory for Particle Physics, known as CERN, proposed a new protocol for the distribution of information. This protocol was based on hypertext—a system of embedding links in text to other text—and this became the World Wide Web in 1991.

Important institutional and regulatory support facilitated the growth. The Internet Society was formed in 1992 as a professional organisation to provide academic and global infrastructure leadership. Key regulatory and trade bodies, such as the Federal Communications Commission (FCC) in Washington, the European Commission (EC), the International Telecommunications Union (ITU) and the World Trade Organization encouraged the Internet to become an international vehicle of 'mass market' communication after being largely oblivious to the international growth of the Internet as a research network in the previous decade. It was largely the academic networks (the National Science Foundation-funded Internet in the United States, Janet in the United Kingdom, and ACS Net, funded by AARNet, in Australia) that negotiated with national carriers to make available the necessary telecommunications circuits and provide inter-city access. It was also the academic networks that stimulated the emergence of the first Internet service providers (ISPs), who provide links between backbone links and Internet customers. However, some de facto US control was maintained through the US government funding of key administrative Internet activities, notably the global coordination of the domain name administration and the global allocations of Internet Protocol (IP) addresses.

Phase Two also saw the Internet emerge as a significant alternative or counterculture mode of expression. The integration of personal computing with international telecommunications connectivity fostered the development of Bulletin Board Systems (BBS), which enabled anyone to become an author and distribute their own messages to anyone else who could also access the Net. Bulletin boards were operated effectively by Chinese students abroad during the Tiananmen Square protests in China in 1989 to take their cause to wide international audiences. Micro communities abound on the Net, covering facets of virtually every aspect of human behaviour and curiosity, with people all over the world searching for content and images about religion, politics, gambling, literature, poetry, drugs, sex and pornography. New kinds of grassroots electronic communities have emerged in ways that would never have been possible through established media institutions.

Phase Three, from the mid-1990s, has seen extraordinary growth of the Internet: in 1981 fewer than 300 computers were linked to the Internet; by 1989 the number stood at 90 000; by 1993 over one million; and by 1996 an astonishing 9.4 million host computers worldwide were estimated to be linked to the Internet, 60 per cent of which were located in the United States, with possibly 40 million people with access around the world.[10] By July 1997 the number of host computers with registered IP addresses had rocketed to 19.54 million.[11] One estimate of the total number of worldwide on-line users in 1999—an educated

Table 5.1 Worldwide on-line users, 1999 (in millions)

Africa	1.14
Asia/Pacific	26.55
Europe	33.71
Middle East	0.78
Canada and United States	87.00
South America	4.50
Total	**153.5**

Source: NUA Internet Surveys, February 1999, at *http://www.nua.net/surveys/how_many_online/ index.htm* guess compiled from various sources—is 153.3 million, with the continental distribution as shown in Table 5.1.

Phase Three has also witnessed the emergence of a commercial marketplace for Internet trading that remains in a highly fluid state. Telecommunications companies have devised aggressive Internet marketing strategies and entered the Internet access business, and ISPs have appeared from nowhere in their hundreds in country after country. IBM and Microsoft changed their corporate strategies between 1991 and 1994 to leverage off the growth of the Internet. The release of Windows 98 in June 1998, with the Microsoft browser integrated into the desktop, showed Bill Gates' strategic commitment to capitalise on the amazing growth of the Internet. This subsequently led to intense legal battles over the issue of whether Microsoft was engaging in anti-competitive practices. Meanwhile, hyperactive newcomer Netscape became the number one darling of the New York stock market, which seemed to defy financial gravity when it came to Internet stocks. A stock market investment of US$1500 in August 1996 in the search engine company Yahoo! would have yielded a staggering return of US$17 000 by April 1998![12]

For many Internet-based business ventures this has been an era in which major stakeholders have been prepared to invest on perceived long-term value rather than seeing impressive immediate rates of return judged by traditional bottom-line performance. If this is a new gold rush—just like the old gold rushes—it is the people selling the picks and shovels who are making money (such as the carriers who provide access to the Net), rather than those who are still searching for the big nuggets of gold.

A *superhypeway*, rather than a superhighway, perhaps.

Cyberspace and virtual communities

The Internet represents a radical departure in terms of the ownership and control of electronic media. Historically, as a result of the limited spectrum space available for broadcasting of radio and especially television, governments have chosen to select particular stakeholders who are licensed to broadcast. Unlike the press, where anyone in Australia can establish and publish a newspaper of their own making, with radio and television an individual, or a group of people, may own a broadcast company but the licence to broadcast is controlled by the Commonwealth of Australia. So, where does the Internet model fit in terms of ownership and control, and how does this affect use, access and content outcomes?

A school of thought emerged in the 1990s, based around the Washington seaboard of the United States, that the Net constitutes a major change in the process of human interaction. Mitch Kapor, designer of Lotus software and founder of the Electronic Frontier Foundation in the United States, argues that the crucial political question is: *'Who controls the switches?'* Users, according to Kapor, 'may have indirect, or limited control, over when, what, why, and from whom they get information and to whom they send it. That's the broadcast model today, and it seems to breed consumerism, passivity, crassness and mediocrity.'[13] However, with the Internet, for Kapor and a significant group of followers, the very nature of the evolution of the Internet has meant that a different model has emerged. He adds, '[U]sers may have decentralised, distributed, direct control over when, what, why and with whom they exchange information. That's the Internet model today, and it seems to breed critical thinking, activism, democracy and quality.'[14] Kapor does not really try to substantiate these bold assertions but stands as one of the prominent members of a cyberspace movement, pioneers and zealots of the computer-mediated communication networks that have flowered in the past decade. John Hindle, Vice President of Nortel, also eulogises that 'the spirit of the Internet is about becoming rather than being, about incubation rather than cultivation, about the myriad possibilities of order rather than any given order, about resilience rather than strength'.[15]

Nicholas Negroponte, cyberspace guru and author of *Being Digital*, also advocates the Internet as representative of an age of transformation. 'Like a force of nature,' Negroponte claims, 'the digital age cannot be denied or stopped. It has four powerful qualities that will result in ultimate triumph: decentralizing, globalizing, harmonizing and empowering.'[16] John Barlow offers even grander assertions, claiming nothing less than 'the promise of a new social space, global and antisovereign, within which anybody, anywhere can express to the rest of humanity whatever he or she believes without fear. There is in these new media a foreshadowing of the intellectual and economic liberty that might undo all the authoritarian powers on earth.'[17]

Contributors to the magazine *Wired* have also brilliantly captured what they feel is the spirit and ethos of this emergent cyberculture. Jon Katz wrote, '[P]robably the first thing anyone notices when they go on-line, however, is the community building taking place all through cyberspace. Old people talk to old people, lonely gay teens find other lonely gay teens, unpublished poets trade with unpublished poets, physicians swap case histories with physicians, parents of dying children comfort parents of other dying children, plumbers order parts from plumbers, truckers chat with truckers.'[18] Howard Rheingold, in *The Virtual Community*, argues that computer-mediated communication can help bring people closer together through their participation in such public conversations and e-mail. He argues that these *virtual communities* foster increasing communicative interactions between people who may be as far as half the world away from one another in either *one-to-one*, *one-to-many* or *many-to-many* contexts. As cynic turned convert, Rheingold cites 'The Well' (Whole Earth 'Lectronic Link') as a San Francisco-based virtual community where people exchange private e-mail and carry on public conversations in more than 260 'conferences' —coffee houses to jazz clubs—where thousands of topics of interest are accessible to people around the world.[19] For Rheingold, real-life interactions are replicated in virtual communities in that people 'exchange pleasantries and argue, engage in intellectual discourse, conduct commerce, exchange knowledge, share emotional support, make plans, brainstorm, gossip, feud, fall in love . . .'.[20] In essence, cyberculturalists argue that this is the grassroots community interactive process in action, diametrically opposed to the centralised top-down broadcast model that has dominated the past 50 years of public communication.

In *Life on the Screen*, Sherry Turkle suggests that we have the opportunity to build new kinds of virtual communities with people from all over the world with whom we may converse daily but never meet, and with whom we may have a fairly intimate relationship. She wrote about Multi Domain Users (MUDs) who create their own texts, and where a computer command links one participant to another computer on which another participant and MUD program also resides. MUD dwellers can go further and create their own identity, inventing themselves as they go along. Turkle writes: '[O]n MUDs, one's body is represented by one's own textual description, so the obese can be slender, the plain beautiful, the "nerdy" sophisticated.'[21] A *New Yorker* cartoon captured the potential for MUDs to experiment with one's construction of identity. In it a dog, with its paw on a computer keyboard, explains to another dog, 'On the Internet, nobody knows you're a dog!'

Among the most fascinating Internet debates are the claims for such reconstruction of identity, or reinvention of one's self, and also for the utopian democratisation prospects that this new form of communication may bring. A new sense of community is emerging, so the argument runs, to form what Rheingold refers to as 'webs of personal relationships in cyberspace', or Turkle's 'sharing of imagined realities', or Kapor's 'extraordinary multi faceted purposes' to which cyberspace is put, or Negroponte's new sense of 'harmony and empowerment', all part of an emergent electronic culture.

Is the Internet a paradigm shift in the sense that it is in the process of redefining the notion of community? The term 'community' is somewhat elusive, though it usually denotes social relationships with common or shared interests, or identity, which operate in specified localities. Fernback and Thompson argue that the notion of community has changed from pre-industrial, to industrial, to post-industrial or information society. In pre-industrial society, they argue, the material component of community was found in the physical, spatial context of a town or province, and the symbolic aspects of community existed in the common linguistic, religious and cultural milieu of the people. The pre-industrial connotation attached to the term 'community' was understood as 'an organic sense of community, fellowship, family and custom, as well as a bonding together by understanding, consensus and language'.[22] The industrial revolution, they add, created inequalities of gender and class in an increasingly divided and isolated population: capitalism had changed people from being interconnected, active and creative beings, to becoming self-interested individuals driven to increase their individual power through wealth. People became alienated from family and meaningful, satisfying work, and institutions reinforced class divisions. Later, the move from an industrial to a post-industrial or information society saw a further renegotiation of the definition of community. Fernback and Thompson suggest that the digitalisation of communications caused us to distance ourselves from our environment in a way that 'abrogated space and time so that we effectively live in a boundless global village'.[23] Hence, in an 'information society' a critical component of community is now found, for those fortunate enough to have access to the Web, in the personal relationships between individuals communicating with each other in cyberspace.

However, before we herald a new era of libertarian self-expression, or of community rediscovered, our critical judgment needs to acknowledge several complex qualifying issues. While there is only limited research on how people actually use the Internet, the predominant mode of use appears to be accessing conventional information offered through a variety of search engines and browsers, particularly by users in educational institutions, corporations and public institutions. Institutional access is usually greater than domestic access, and although the Internet shows impressive rates of take-up around the world, especially since the mid-1990s, users do need to have access to a personal computer, a telecommunications network, generally a modem for domestic access, and need a relatively high level of computer literacy. The hardware/software skill set needed for this new medium is very different from those required for other communications technologies, notably the telephone and television, which are much more readily available and easier to use.

Just as the coming of the Internet has seen a new breed of 'technological utopians' or 'technological fresh start' theorists emerge, so too have 'technological sceptics' condemned the 'netizens'. There is a school of thought that has come to question the alleged assault on 'reality' by men, women and children who are heavy Internet users, alleging that these people have become more and more withdrawn from society and more insular: as Paul Virilio views it, 'a society of cocoons . . . where people hide away at home, linked into communication networks, inert . . .'.[24] Mark Slouka of Columbia University argues that the Net's virtual reality reduces our options and strips us of our humanity and the capacity to interact with the outside world.[25] Clifford Stoll suggests that 'computer displays only weakly imitate the sounds, sights, smells, tastes, and touches of nature'.[26] The growth of the Internet may lead to a revival of 'Life—Be In It' campaigns which remind us that we can make plenty of choices about how we organise our lives.

United States academic researchers have recently explored issues about the alleged personal isolation and alienation of Net users. Systematic research was undertaken by a group of psychologists at the Carnegie Mellon University who used longitudinal data to examine the effects of the Internet on social involvement and psychological well-being.[27] Their HomeNet study consisted of a sample of 93 families from eight diverse neighbourhoods in Pittsburgh, Pennsylvania. They argued that 'the findings of this research provide a surprisingly consistent picture of the consequences of using the Internet. Greater use of the Internet was associated with small but statistically significant declines in social involvement as measured by communication within the family and the size of people's local social networks, and with increases in loneliness, a psychological state associated with social involvement. Greater use of the Internet was also associated with increases in depression.'[28] While this is pioneering exploratory research, its authors conclude by recommending that 'people should moderate how much they use the Internet and monitor the uses to which they put it'. This psychological territory is likely to become a growing research field in the next few years.

We have become more dependent on abstract forms of communication in a world where spectatorship, or voyeurism, has increasingly replaced direct first-hand experience. *Cyberspace alters our notions of space from physical space to transactional space; VR, virtual reality, alters our sensory perception away from RL, real life.* In cultural terms, convergence has blurred our boundaries between reality and illusion, between our direct sensual world and illusion, and between the local and the global, potentially redefining our inner self.

Do these changes, collectively, benefit us? We will need to reassert our sense of inner self in this virtual age. The real body will forever remain disjointed from the virtual body. There are, of course, choices for human beings in the ways they can construct their lives: we can still go for a walk, enjoy jokes and talk to each other directly. We can still reject a life based on virtualism, on-line shopping, banking and gambling, 500 television channels, Net surfing and computer games. Like it or not, though, it is the way the new communications world is heading. Ken Karakotsios, former Apple employee, sums up the scepticism thus: '[T]he only thing wrong with the universe is that it is currently running someone else's program.'[29]

Changing place and space

The nature of the new media involves major changes in the processes of communication and global impacts on society. The community of the Internet has changed theoretical notions about space and place in our lives. This was predicted by a succession of communication scholars who have examined the effects of new media, particularly the growing importance of the more indirect human relationships that new media bring, technologically enabled by computers and satellites. These include Arthur C. Clarke or Marshall McLuhan's notion of a 'global village', changing spatial relationships via new media, Harold Innis's 'time binding' and 'space binding' technologies, and Anthony Giddens' 'time–space distantiation'. Writers of globalisation refer to the 'compression' of the world, through global television forms, while cyberspace zealots tend to refer to the 'stretching' of our cultural boundaries that the Internet makes possible. The term 'cyberspace' implies taking one into a created space that previously did not exist, into the ether of global 'any-to-any connectivity' or into 'transactional space'. New forms of multimedia also oblige users to reconsider the form of reality of experiences when they feel taken out of their direct sensory selves with 'virtual reality'.

Established media forms, notably broadcasting, are in transition, and there is now the prospect that Marshall McLuhan's 1960s' notion of a global village may be finally realised in some key senses. There are real prospects that the new media will lead to the creation of much more open and accountable societies. Today any major trouble spot in the world sees the world's communications apparatus quickly focused upon it to distribute its brand of instant analysis more widely than ever before. Take, for instance, the volume of world media attention that converged on China's Tiananmen Square in 1989, the events that led to the downfall of Ferninand Marcos in the Philippines in 1986, and those leading to the resignation of President Suharto in Indonesia in 1998. Ruling elites, wherever they might be located, will find it harder to hide from world attention and get their military to crush popular uprisings with complete disregard for human rights.

The student-led revolt in Indonesia in mid-1998 which led to the overthrow of President Suharto encapsulates some of these notions. Students initially protested in the social space of their own geography across their campuses. When they moved to the streets and later occupied the grounds of the Parliament, the space of the nation state, they attracted the world's global television networks. The gladiatorial contest went out to every newsroom in the world, via satellites, as the day's leading international story. There was, for just a few weeks, intense and uniquely open analysis by a barrage of international media people of the circumstances which led to the downfall of a President who had ruled the world's fourth most populated nation for more than 30 years. Use of the Internet guaranteed that grievances and pressures for change would be depicted, as Indonesian protesters e-mailed their version of the causes and events on the World Wide Web to influence people in other parts of the world. And again in Malaysia in September 1998, when Malaysian Prime Minister Mahathir Mohamad sacked his deputy, Anwar Ibrahim, supporters of Anwar rallied to get tens of thousands at street protests—via the Internet. This is a unique form of the use of communications in the history of uprisings. Reform movements can now use the transactional space of the uncensored Net, a medium much more constructive than that of the journalism of the daily newspapers where local journalists often toe the state line and report only what the ruling elite want to be reported. No longer, it seems, can isolation be achieved by shutting down the few daily newspapers and taking out the radio and television transmitters. *There is, as Anthony Smith has said, the potential for 'a new moral order'.*[30]

We might do well to look also at Western examples in this context. The Net coverage of the events surrounding President Bill Clinton's affair with a White House intern has been regarded as a landmark in political participation and discourse. Thirteen months after Internet gossip columnist Matt Drudge put the Clinton–Lewinsky affair on the World Wide Web, the President did a backflip after initial denials and said in the White House, 'I am profoundly sorry for what I said and did.' Americans devoured Special Prosecutor Kenneth Starr's report on the Net; the twenty million Americans who accessed it is the highest number of people ever to access a single document on the Internet. Starr's extraordinarily explicit report of this sexual affair was there for all netizens to read, and it put Clinton's presidency, and Monica's future personal life, under enormous pressure. Thirty years ago, philanderer John Kennedy, one of Clinton's presidential predecessors, was not subject to anywhere near this extraordinary scrutiny and publication of matters relating to his intimate personal life. Do we still need censors to draw lines?

Governance of the Internet: Censorship and privacy

The Internet is widely perceived as the most open and 'anarchic' communications platform ever, liberating its users from so many of the usual restrictions imposed upon conventional publishing and broadcasting, and flowering a new era of self-expression through virtual communities. The other side of this new libertarian era is a set of complex issues about perceived problems with some Internet content. The Australian Competition and Consumer Commission said in 1997 that their cursory examination indicated that at least 300 Australian web sites raised clear concerns about possible breaches of the *Trade Practices Act*.[31] Moreover, there is a myriad of concerns about allegedly damaging or undesirable content that the Net has fostered: bomb recipes which children can access on the Net; vicious racist hate mail; intimate personal matters made available on the Net from court files; the multi-Net listing of hundreds of people's credit card numbers without their consent; spamming or junk e-mail distributed to users who never requested it; and the proliferation of highly pornographic web sites.

What do we do about these issues or problems, if anything? Underpinning them are bigger questions about governance—of how the world of the Internet could or should be governed. As we have seen earlier, the Internet evolved by design as a direct outcome of the *lack* of governance, and its history of unregulation and freedom of content is widely perceived as its greatest strength.

We cherish our democratic rights of freedom of expression. Australian society has a commendably open door attitude on what we can read in newspapers, see on film and television, and hear on radio. However, there have long been calls for content censorship of many forms, from several quarters with different interests, and Internet content could hardly escape such scrutiny.

If there is any one single area which has been damaging to the reputation of the Internet, and which has led governments to look at possible ways of intervention, it is the haven that the Internet can allegedly provide for paedophiles. Bulletin boards on the Net have fostered networking between paedophiles, enabling them to become more anonymous, less risky and possibly more active. A National Crimes Authority report (1995) said that paedophiles had turned to the Net because the exchange of pornography and names of victims via on-line services was relatively anonymous: exchange was easily achieved, material could be received immediately, and there was unlimited potential to reproduce and distribute it.[32] Paedophiles can be put in touch with young children via fee for service computer mail boxes, clouding the real identity of the source and making detection difficult. There are many reports that those with deviant intentions have been quick to recognise the potential of the Net by offering competition prizes and gifts to gain information about children and their families.[33]

Do we need to find a way of protecting children from predators on the Net? Who would be ultimately responsible for such protection, and how could it possibly be administered? The six-volume report of the Royal Commission into the New South Wales Police Service devotes Volume V to The Paedophile Enquiry, and recommends that the government provide an effective regulatory system, in conjunction with parental guidance and the filtering and blocking of the technology available. On the issue of general content censorship, the Australian Law Reform Commission has argued that 'the basis of present censorship policy is that adults in a free society ought to be able to see what they wish so long as that right does not infringe on the rights of others to avoid offensive material, and adequate protection is given to children. Current policy sees offensiveness in terms of sex and violence, and particularly, any combination of the two.'[34]

The regulatory body currently responsible for censorship in Australia is the Office of Film and Literature Classification, a semi-independent government agency under the jurisdiction of the Commonwealth Attorney General, which administers the *Classification (Publications, Films, and Computer Games) Act 1995*. This type of regulation does not easily translate to the Internet because, whereas a film deemed to be unacceptable for screening by the above authority can be withdrawn by its distributors, or a computer game banned from retail stores, with the Net no entity controls its operation, or any of its content. The censorship lobby cannot claim that there are no controls at all on how material is accessed. 'Adult' material does not accidentally come across the Net—it has to be deliberately accessed. The traditional argument against content censorship is that bans by government regulators only serve to drive the problem underground, making the product more attractive and expensive on a black market.

Governments around the world are all grappling with the problem of how to monitor and regulate illegal and unacceptable content on the Internet. In the United States, following concern at the amount of 'objectionable material' available on the Internet, the government devised the *Communications Decency Act 1996*. The intention of the legislation was to make it illegal to put 'indecent' or 'patently offensive' words or pictures on the Internet where a person under eighteen would be able to access it. However, the American Civil Liberties Union and the American Library Association took civil action against Janet Reno, US Attorney General, and the US Department of Justice on the grounds that the Act infringed the first amendment, and was thereby unconstitutional. Judge Slovitor ruled without hesitation that the Act was unconstitutional.[35] He argued that he accepted without reservation that the government has a compelling interest in protecting children from pornography but that this can be done 'through vigorous enforcement of existing laws criminalizing obscenity and child pornography'. He concluded, '[T]he Internet may be regarded as a never ending world wide conversation. As the most participatory form of mass speech yet developed, the Internet deserves the highest protection from government intrusion.'[36]

This is not the position taken by many other nations. In China, for instance, all computer links to the Internet are controlled by the Ministry of Posts and Telecommunications, all Internet access providers must be approved by the government, and all Internet users in the country must register with security forces within 30 days of obtaining Internet accounts. A declaration is required by creators of web sites that no criticism of the government of China will be put on the Net. The policy is to tightly control the country in terms of 'desired' information flow.[37] Singapore, although encouraging the development of the Internet, has 'established a class licence system for Internet service providers and content providers, regulating the conduct of members of the industry and also the on-line content'.[38] The three Internet service providers in Singapore (Australia has 800) can be required to remove the whole or part of a program included in its services once the Singapore Broadcast Authority has informed the licensees that the content is contrary to their Internet Code of Practice, or that the broadcast is 'against the public interest, public order, or national harmony, or offends against good taste or decency'.[39]

Neighbouring Malaysia, with its Multi-Media Super Corridor plans, regulates by applying conditions of service upon its only two Internet service providers, and customers are bound by conditions of access; their service is terminated if they fail to comply. The Malaysian government has also established two e-mail hot lines for users to report any illegal activity. It will be fascinating to watch how other governments during the next few years respond to the complexities of the growing body of on-line content services in terms of their future regulatory stance.

For some time the Australian approach in this context was essentially to develop a regulatory regime based on shared responsibility between industry, government and the community. Rosalie O'Neale, of the Australian Broadcasting Authority, points out that the ABA has taken a middle ground position because 'simple solutions based on national and centralised management of content are no longer appropriate or workable', but that 'self regulation is unlikely to provide the complete solution either'.[40]

Any attempt to replicate the broadcasting content regulation model faces the fundamental problem that Internet service providers can hardly be expected to have prior knowledge of the content being disseminated via their service. In the view of the ABA, argues O'Neale, *a co-regulatory regime*, within which government, industry, law enforcement agencies, community groups, educators and parents each play a role is the only way that this balance can be achieved'.[41]

In June 1996, the ABA's report, *Investigation into the Content of On-Line Services: Report to the Minister for Communications and the Arts*, recommended a regulatory framework for on-line service providers based on codes of conduct developed and administered by the on-line services industry. This pioneering report advocated the development of international collaborative arrangements for Platform for Internet Content Selection (PICS), which provides an infrastructure for content labelling as site 'ratings', allowing for a user ratings system rather than outright censorship of Net material. The rating enables the user to know in advance from its rating category the nature of the particular material which could be called up on screen, thereby providing the opportunity for user-based censorship. PICS itself does not provide a form of censorship, merely a framework in which judgments can also be made by any organisation, such as a government department, or an official censorship body or individual user censorship.

Following the ABA report, the Commonwealth Minister for Communications and the Arts, and the Attorney General, announced in July 1997 the regulatory framework for on-line services in Australia for public comment. The overriding theme which underpinned these

47 principles was that 'material accessed through online services should not be subject to a more onerous regulatory framework than "offline" material such as books, videos, films and computer games'.[42]

The major legislative principles were:

- facilitation of the codes of practice by the on-line service provider industry to facilitate effective, appropriate and fast complaints procedures;
- that the ABA will be the new regulator of the scheme, requiring a substantive amendment to the *Broadcast Services Act 1992* which did not legislate jurisdiction of the ABA over the Internet;
- to encourage the states to support the national industry-based scheme;
- to pursue the development of international collaborative arrangements, notably the Platform for Internet Content Selection; and
- to encourage greater community awareness and foster education programs.[43]

Rating and labelling are gathering more attention throughout the world as a way of dealing with Internet content which is considered harmful to children. US President Bill Clinton expressed his support for the use of content rating and labelling schemes, suggesting that 'we need to encourage every Internet site, whether or not it has material which is harmful for young people, to label its own content'.[44]

Politicians find themselves under great pressure from members of their electorates who occasionally become outraged by something they regard as highly offensive that has appeared on the Net. The Australian government wants to be seen to be responsive to such protest on the one hand, yet on the other hand not become dictatorial in stifling freedom of expression. The general Australian policy principle of 'hands off' the Internet has more recently shifted towards more draconian attempts at censorship. State and federal Attorneys General, meeting in Hobart in December 1997, foreshadowed criminal sanctions for individuals who place 'offensive or illegal material on the Internet' under tough measures.[45] How, though, could legislation define 'offensive' material? And who would ever be able to act against whom—the Internet service provider, the writer or the user?

As the pressures mounted on government to censor particular Internet sites, especially from parents concerned about how easily children could access pornography, the government moved to block access to certain material. The Howard government initiated the *Broadcasting Services Amendment (On-line Services) Act* which passed the Parliament in May 1999. The Act established that the Australian

Broadcasting Authority is responsible for filtering access to categories of 'Refused Classification', 'X' and 'R' rated material. Under this legislation the ABA can order an Internet service provider to remove offending material from the Net within 24 hours and failure to do so could lead them to be fined up to $27 000. Michael Ward, Vice President of OzEmail, attacked the legislation and argued that 'there are no reasonable steps that are technically feasible or commercially viable to block that material . . . short of creating a separate Internet in Australia with a filter between us and the rest of the world, there is no way of doing this'.[46] The Howard government subsequently faced widespread condemnation for passing this legislation.

Ultimately, it may be much better to 'err' on the side of freedom of expression and self-censorship than to propose such modes of external interference or unworkable attempts at content control for the Internet.

New capitalism of convergence: The Internet economy

The Internet has evolved in several phases, progressively centred on defence, research and academia, and its shift into commercialisation in the late 1990s raises the issue of whether a new trading paradigm is now emerging that may become an Internet economy. The term 'electronic commerce' (e-Commerce) is most commonly used in this context and this may be defined as *using the Internet to promote or sell products or services*.[47] The term is also sometimes used to encompass other forms of network-based commercial transactions, such as Electronic Funds Transfer at Point of Sale (EFTPOS), commonly used by consumers with credit cards in their retail transactions, or Electronic Document Interworking (EDI), used to send sales orders, invoices and confirmatory documents electronically in a corporate supply chain, but this section of the book will deal only with Internet-based trading.

Internet sites are now rapidly emerging throughout the world, marketing an innovative range of goods and services with often unfamiliar or experimental brands in an unprecedented phase of global commerce. Consumers with Net access can now buy books, CD-ROMs, jeans, suits, groceries, travel services, video and audio, greeting cards, wine, computer software, stocks and shares, as well as gamble offshore or process their personal banking transactions on-line, if they wish. This, however, is new territory for business in terms of the creation of an Internet with its associated marketing, distribution, consumer transactions and other interchanges with customers. At this early stage of commercialisation the most critical factor in the success of Net trading appears to be consumer perception of value, such that where consumers can buy goods and services cheaper via the Net, or where the cost of transactions is significantly lower than by conventional means, the signs are promising. Yet as *The Economist* reported, '[D]espite the Internet's rapid growth, it is still producing more hype than money . . . Many Internet start-ups have run through their venture capital and out of money.'[48]

Electronic commerce takes on many forms in its evolution, which may be summarised as follows:

- *The retailing of goods to consumers*, such as books, CD-ROMs and clothes. This is still in its formative stage and Internet consumer product marketing has not reached the so-called critical mass which conventional retailers and investors regard as vital to run profitable businesses.
- *Business-to-business transactions.* This involves transactions between companies for goods and services, such as manufacturers who automate their ordering with their suppliers using the Internet, and is widely predicted to be the biggest growth sector of Net trading in the next five years.
- *Internet service transactions.* Typical services here include financial services, professional consulting, and on-line education and training courses. Financial services are starting to take off, but acceptance of on-line education generally remains low.
- *New marketing groups.* Several established brand names, such as Amway and Tupperware, which traditionally used interpersonal group marketing techniques to build highly successful marketing empires, are looking at using the Net as a new mode of marketing their products.

It is difficult to estimate accurately the dollar value of the e-Commerce economy, internationally and in Australia, although there is no absence of consultants who offer such estimates as well as advice on how to get into this business. Andersen Consulting estimated e-Commerce in Australia at $85 million in 1997, The International Data Corporation's (IDC) figure was $127.3 million (1997), but the Department of Industry, Science and Tourism valued on-line purchasing more conservatively at $55 million.[49] The median figure for e-Commerce in Australia in 1997 amounted to a mere 0.02 per cent of GDP. Meanwhile, estimates of the value of worldwide Internet commerce in 1997 range from IDC's US$2.6 billion, to a US Congress report at US$7 billion, to an Andersen

Consulting report at about the same, A$10.5 billion.[50] Clearly, the United States is the international market leader in electronic commerce.

Estimates for the growth of e-Commerce, from both public and private sources, all postulate significant increases in the next few years, although again there are significant variations among the e-Commerce futurists. Estimates for the world growth of e-Commerce from 2000 to 2002 range from that by the Australian Department of Foreign Affairs and Trade at US$100–150 billion, to the US Congress at US$300 billion, to IDC's US$333 billion.[51] Most analysts agree that the success of Internet electronic commerce lies in the business-to-business market sector. Forrester Research estimates that by 2002, business-to-business transactions in Australia will ramp up substantially to A$327 million, which would be a significant shift in Australian business practices.[52] Estimated growth rates in Australia for business-to-consumer activities are lower. Products such as books, music and software, where consumers tend to have a high level of product familiarity and can often buy cheaper on the Net, are the most successful retail e-Commerce items in Australia. Other shopping appears to remain tied to established behavioural habits, and home shopping for perishables has not reached any significant take-up. Home banking is growing, although it has not yet achieved substantial numbers because of data protection issues and some lack of consumer confidence.

There are some important signs in the United States, however, that e-Commerce may well constitute a sea change of business and consumer practices during the next few years. Dell now sells US$5 million worth of computers a day on its web site, claiming that its on-line sales give it a 6 per cent profit advantage over its competitors.[53] Discount-mortgage broker American Finance and Investment, which conducts 60 per cent of its business on-line, became profitable within 90 days of beginning Net trading.[54] Eddie Bauer, an American clothing retailer, has an on-line operation which became profitable in 1997 after revenue growth of 300–500 per cent per annum.[55] Federal Express in the United States saves about US$1 million a month by enabling its customers to track the progress of their packages on-line themselves rather than by labour-intensive telephone inquiries. These are select American examples, but it must be remembered that the potential of Internet trading is global, and that there are opportunities for many businesses, independent of their distance from customers and irrespective of their location. Given that the majority of private purchases by Australians over the Net are made using overseas sites, especially American, the issues of loss of opportunity to date and the potential balance of trade impact for Australia is serious and will be addressed in Chapter 8.

Amazon.com

One of the most remarkable success stories of electronic commerce is that of an American-based on-line retailer, Amazon.com, which now sells more than one million books a day under its slogan 'Earth's Biggest Book Store'. Amazon was founded in 1994 by former Wall Street trader Jeffrey Bezos who recognised that an inherent problem in traditional book retailing is that there are too many books from too many publishers to be carried by any single bookstore, and that the costs of carrying high inventory are crippling for retailers. The Seattle-based company's web site (*http://www.amazon.com/*) was established in 1995 to allow Internet users to browse through their database for titles and to purchase their books on-line. Within a month Amazon.com had attracted customers from 66 countries (now 150), its revenue base was US$147.8 million by 1997, an astonishing 841 per cent growth on its previous year, and for 1998 its total revenue base reached US$610 million.[56] Although in terms of revenue Amazon.com is now surely the most successful retailer in the Internet's short retail history, it has also incurred significant losses, and as at March 1998 its accumulated deficit was US$42.9 million. This is a company which has yet to return a profit to its investors, and in a filing in March 1998 the company advised the US Securities and Exchange Commission that it did not anticipate it would become profitable in the near future! Its stock market capitalisation, however, is now more than that of the total aggregated value of every traditional bookstore in the United States. In the summer of 1997 it took 58 days for trading volume in amazon.com to equal the book retailers' entire shares outstanding. This figure had dropped to 22 days by August 1998 and to thirteen days by December 1998 when its shares rose from US$318.75 on the Monday before Christmas Day, up US$32.063 from the previous Friday.[57] In Amazon.com we see the new capitalism of convergence where the perceived value of its innovation and initial capture of market share far outweighs traditional financial practices where businesses are primarily assessed on their price–earnings ratios. This superbly innovative on-line company is 'riding the wave'.

There are several key ingredients in this new e-Commerce paradigm. Whereas traditional booksellers tend to tie up their funds in large stock inventories, Amazon.com, until recently, did not physically stock any of the books it sold. Instead, they offered customers on-line access to databases from major book distributors which enabled Amazon clients to search for books they want by author, title or subject. The 'search results' screen, bearing the Amazon.com logo, provides the customer with full publishing details, price and estimated shipping times for overseas clients. Customers pre-pay their orders using the company's on-line transaction system. Meanwhile, Amazon itself paid its suppliers within an average of 46 days from the date of ordering, giving it a great cash-flow advantage over its competitors. Moreover, Amazon has another sales edge because it discounts the retail price of its titles between 10 per cent and 40 per cent, and this may be the most critical factor of all in its great success.

Throughout 1998 Amazon began expanding into new markets and diversifying its product range, and announced on 27 April 1998 that it had acquired three Internet companies:

- Bookpages (http://www.bookpages.co.uk/), a British-based on-line bookstore with access to all 1.2 million UK book titles currently in print.
- Telebook (http://www.telebuch.de/), a German-based on-line bookstore providing access to nearly 400 000 German-language titles.
- The Internet Movie Database (http://www.imdb.com/), an on-line database providing capsule reviews of thousands of movies, as well as entertainment industry updates.

These moves show a new arm to Amazon's strategy of shifting into new book-buying markets, particularly non-English-speaking markets, by providing localised versions of its core business. In May 1999 the company announced it had agreed to purchase three more companies—Exchange.com (hard-to-find, antiquarian and used books, as well as music memorabilia), Alexa.com (a web navigation site) and Accept.com (an e-Commerce solutions business). Another factor in Amazon's success is that it cleverly fed off the Internet's 'community building' ethos by encouraging its customers to submit their own reviews of books they had read which were then featured in Amazon's inventory. Amazon has adapted one of the oldest commercial enterprises in the world, selling the books of a 500-year-old medium, to a new medium of global distribution. Amazon has shown that a workable, user-friendly model can be found for electronic commerce.

It has also found, together with its search engine counterparts, a new way of being assessed on the stock market. The case of Amazon.com raises important questions about the way Australia conducts business in the future. What do major Australian booksellers do when they find that a new competitor, with new marketing and distribution practices, is growing at 30 per cent per annum and eating into their markets? Moreover, if new Net trading commercial practices eventually extend into a wide range of goods and services, and Australian businesses do not have a significant Web presence, will this have a major effect on our national balance of payments?

Australian web innovation

To date, Australian business has tended to 'dabble' with web sites. Marc Phillips' book, *Successful E-Commerce*, highlighted case studies of small Australian businesses which have embarked upon creating their own trading web sites.[58] Examples include:

- the famous R.M. Williams with its Boots Online (*http://www.bootsonline.com.au/*), which boasts of taking Australian-made boots to the world and has 95 per cent of its customers overseas;
- Cyberhorse (*http://www.cyberhorse.net.au/*), with its virtual form guide which gets 400 000 hits per month;
- Lovittools (*http://www.lovittools.com.au/IT/FPSite.MTM*), a medium-sized precision engineering company which found that the capacity to transmit drawings, photos and documents over the Net assisted its relationship with customers;
- Netpsych (*http://www.netpsychnet.au/logon.html*), which is used as a vehicle to market self-help psychological manuals;
- Stamps.au (*http://www.stamps.au.com.product.asp?sku=90268%EO*), which offers Internet shoppers the entire range of Australian stamps and philatelic on-line services; and
- The Mail Service (*http://www.lotteries.net.au/*), which offers Lottery products on-line and targets the Pacific Islands for customers.

Minister for Communications, Richard Alston, has issued a report card of successful case histories of e-Commerce. These include Absolut Beach (a swimwear shopfront on the Web), Green-Grocer (specialising in fruit and vegetable delivery), Oz Coffee (on-line suppliers of coffee, equipment and accessories) and Webloans (on-line mortgage brokering services).

There are many such small-scale creative initiatives now emerging in the on-line world, although in terms of generating revenue and traditional business bottom-line performance they have a long way to go. There is as yet no Australian equivalent to amazon.com in terms of building a retail brand name, a substantial electronic commerce customer base with huge revenue growth and a brilliant share value. The most commercially successful small Australian Internet company has been OzEmail, an Internet service provider, not a Web retailer. This was a fledgling R&D company in 1992 experimenting in Internet-related software which Kerry Packer, in a rare moment of bad commercial judgment, virtually gave away. The company was acquired by Malcolm Turnbull (republicanist and merchant banker), Trevor Kennedy (former Managing Director of Packer's Australian Consolidated Press) and Sean Howard (former magazine publisher). They built OzEmail as a successful Internet service provider, and the market came to be dominated by Telstra's Big Pond and OzEmail, each of which shared about 20 per cent in 1998 of the booming Internet service market followed by a long tail of battling service providers. But like most Australian home-grown successful IT ventures, the company attracted the eagle eye of foreign interests, and MCI WorldCom successfully bid $520 million in December 1998 to acquire OzEmail.

For the past couple of years, the question has been asked whether a new economy is evolving around the Internet. The American economy in the late 1990s has been held together to a large extent by the extraordinary boom in high-technology stocks and shares, in an economy where traditional manufactured goods have become more uncompetitive due to the consistently high value of the US dollar. Microsoft stock bought at a value of US$10 000 in 1989 was worth US$250 000 early in 1999. The significant new pocket of booming stock market valuations in recent years has been the Internet stocks. Bloomberg's reported in December 1998 some amazing returns on investment for twelve months of the 'Nifty Nets', to use the trade jargon—Amazon.com's growth was 1080 per cent, yet it has never made a profit, followed by Yahoo! (700 per cent), America Online (452 per cent), and Onsale (403 per cent).[59] The US Internet Index almost quintupled in the year preceding the end of the first quarter, 1998. However, Microsoft Chairman, Bill Gates, warned early in 1999: 'I don't recommend Internet stocks to people who don't like massive risks, especially at current prices. People are jumping into it as though it's a gold rush.' (See Michael Wolff's book, *Burn Rate: How I Survived the Gold Rush Years on the Internet*, Weidenfeld & Nicolson, London, 1998.) Similarly, US Federal Reserve Chairman, Alan Greenspan, argued in February 1999 that although much of the present form of the distribution of goods and services would 'eventually move into some form of Internet system', he warned that 'the vast majority of small Internet companies are almost sure to fail'.[60]

There appears to be a speculative financial hierarchy here. There are the established players, such as Microsoft, who own the software that has become the world's ubiquitous computing platform, or the telecommunications carriers, most of whom are likely to be in business forever. Then there are the productive Internet innovators who have built new kinds of businesses although they are unrealistically valued on the stock market, such as amazon.com which grew by 1080 per cent in one year (1998), but which has an 8.4 million customer database and a high level of customer loyalty. Then there is a category of seemingly crazy operators who have created some kind of Internet company which has no apparent development foundation and whose share market price appears to defy any conventional investment logic. The American stock market has become overheated by the Internet speculators—hordes of investors searching to get in on the ground floor of the stock that may become the next Microsoft. Though there are occasional market corrections for Internet stocks in the United States, such as in July 1999, one fear is that a major Internet stock tumble might take the whole of the market with it. In a world where Japan, much of South-East Asia, Russia and Latin America are economically unstable, and in a period when some economies have been on the brink of collapse, there is a sense of profound unease that unprecedented and unrealistic high American share market valuations are critical to the overall health of the global economy.

So, where does Australia stand in terms of an emergent Internet-based new economy? To date, Australian business has tended to adopt a 'wait and see' attitude to e-Commerce. Research in Australia by Andersen Consulting in April 1998 indicated that:

- Eighty per cent of CEO respondents and 95 per cent of government respondents agree that e-Commerce will 'revolutionise the way they do business in five years'.
- Most CEOs do not presently regard e-Commerce as one of their highest priorities. For 81 per cent of respondents, e-Commerce does not appear in their top four strategic priorities.
- Fourteen per cent of respondents 'have a comprehensive and well developed strategy in place' and 64 per cent 'are either developing or near completion of a strategy'.[61]

Peter Gerrand, CEO of Melbourne IT, points to electronic commerce beginning to move in Australia from its existing small base. He reported in October 1998 that:

- some 74 per cent of Australian businesses planned to have Internet access by the end of 1998;
- only 7 per cent of Australia's 900 000 businesses have distinct web sites—characterised by distinct Internet domain names, such as *qantas.com.au/*; but that
- more than *2500 Australian businesses per month are getting their domain names registered.*[62]

A paradigm shift?

The Internet does represent a paradigm shift—Kuhn's 'universally recognised scientific achievement' with 'all sorts of problems for adherent practitioners to resolve'—in the world of contemporary communications. What a remarkable new communications platform is the Internet! From many residences and businesses in Australia, useful information can be accessed via the Internet from more than three million web sites in more than 150 countries. The Internet has developed extraordinarily rapidly from its specialised institutional American-funded defence origins of the late 1960s, into a more open, ideologically diverse, community-based communications platform in the world of the 1990s. International telephony networks remain more accessible to more people, but they do not offer the richness of form as a medium of communication of the World Wide Web. The Internet is still, however, predominantly an English-based medium, and it may be many years before even 10 per cent of the world's population have regular access to it.

Regarding electronic commerce, the sceptics argue that the highly inflated US stock market value, compared with the real price–earnings ratios of many of the search engine companies, and the perceived value of a company such as the innovative on-line bookseller amazon.com, are evidence of 'oversell', or a false dawn for the notion of an Internet economy. In December 1998 the stock market capitalisation of the search engine Yahoo!, with few tangible assets other than a powerful search engine, some software and a good brand name, was roughly equivalent to that of the 'big Australian', BHP, with its petroleum, natural gas, steelworks and many energy fields, and a long history of strong price earnings. The US economy has been boosted in recent years by its high-tech stocks, the perception of their long-term value, and by investors' dreams that they will find another Microsoft. Worth remembering, though, even if there are some dives and crashes on the way, is that this new phase of commercialisation of the Net is underpinned by a new communications platform—the World Wide Web —and by a staggering growth of take-up throughout the world today. Companies and investors come and go, it seems, but communication infrastructures live forever.

An inherent strength of the Internet is its anarchy compared to the established modes of ownership and control of traditional media: there are no direct equivalents to the 'gatekeepers' of content and form which characterised the major media of the past few decades, the press and broadcasting. Everyone who has access to the Net can become their own author, expressing their own sense of identity to other Net users scattered throughout the world. Herein, though, lies the predicament for the future of the Internet: given its growth and business potential, will it be captured by commercial interests, in several contexts, during the next decade?

The acceptance of the Internet by the community in recent years raises a most fundamental question of whether we are witnessing the emergence of new ways by which individuals communicate with each other, a paradigm shift in the communication process. Are those who access the Net—netizens, as they have been called—members of a new communications culture? Jon Katz's classic article, 'Birth of a digital nation', suggests 'the primordial stirrings of a new kind of nation', which is also articulated elsewhere as the emergence of an electronic or digital culture, flowered by the range of creative manifestations of the Internet. Katz writes, '[T]his nascent ideology, fuzzy and difficult to define, suggests a blend of some of the best values rescued from the tired old dogmas—the humanitarianism of liberalism, the economic opportunity of conservatism, plus a strong sense of personal responsibility and a passion for freedom.'[63] Grand sentiments from a member of the elite of the richest country in the world, but unlikely to be realised by so many people throughout the world who may never experience access to the Internet. Thomas Kuhn argued that when a revolution is occurring there is a period of considerable overlap and continuity of paradigms rather than a dramatic or sudden shift. At this stage in its evolution the Internet offers both promise and predicament as the dawning of a new communications era.

It is within cultural terms that ultimately we will assess whether the Web of connectedness will eventually evolve as a global new communications order—as a cultural paradigm shift.

6 Being human: Paradoxes of the new media

> *New technologies alter the structure of our interests: the things we think about. They alter the character of our symbols: the things we think with. And they alter the nature of community: the arena in which thoughts develop.*
>
> Neil Postman, *Technopoly*, Vintage, New York, 1992, p. 20

> *I do not want my house to be walled in all sides and my windows to be stuffed. I want the cultures of all the lands to be blown about my house as freely as possible. But I refuse to be blown off my feet by any.*
>
> Mahatma Gandhi, quoted in *Opening Windows*,
> AMIC, Singapore, 1996

We live in a radically changing society where major forces are challenging our outlook and altering our identity. Many of the secure, familiar pillars of our society are in a state of transition, such as the nature of work, the security of long-term employment and our sense of identity based on occupation. We are witnessing the decline of the traditional nuclear family, organised religion and trade unions. No longer do we tend to see ourselves as citizens of stable neighbourhoods with friendly local businesses and healthy community-based activities. There appears to be common acceptance that there is little choice other than to work within the macroeconomic parameters set by the global economy, and find ways of maximising the new opportunities that come with globalisation. A sense of powerlessness is now common, where so many people feel loss of control over their jobs, their choice of livelihood, their environment and much else in their lives. A principal agent driving these complex changes is the increasing infiltration of new communications technologies.

The notion of an information society masks a series of unresolved complexities about the nature of societal changes, and who benefits. The prime characteristic of the information revolution and its impact is, oddly, its ambiguity. Why have we embarked upon such radical changes we know so little about, and which bring so many major contradictions? This is an era of unprecedented growth and wealth for communications, yet major job growth has not followed, mega-company alliances are increasing, yet so too are the number of small service organisations, and many institutions are fundamentally re-evaluating their role in society.

A central challenge to our notion of identity is emerging from contemporary redefinitions of work, occupation and employment. Most 'post-industrial society' theorists of the 1970s and 1980s correctly predicted the significant infiltration of information technology in industrial practices, and the growth of a service sector in modern economies, but their predictions of a post-industrial utopia for work and leisure have been sadly unrealised. In 1980, Peter Large wrote in *The Micro Revolution* that 'this quiet revolution has produced a few horrors and many niggling annoyances, caused by inadequately planned computer systems and the scarcity of computer-skilled people'.[1] Christopher Evans, in *The Mighty Micro*, similarly claimed that 'the new unemployment, if we can use the dreaded word at all in this context, will be of quite a different order, and much closer to a kind of affluent redundancy'.[2] British academic Tom Stonier's book, *The Wealth of Information: A Profile of the Post-Industrial Economy*, gallantly concluded in 1983:

> [J]ust as industrial economy eliminated slavery, famine and pestilence, so will the post-industrial economy eliminate authoritarianism, war and strife. For the first time in history the rate at which we solve problems will exceed the rate at which they appear. This will leave us to the real business of the next century. To take care of each other. To fathom out what it means to be human.[3]

Grand ideals indeed for our post-industrial society.

Later, German authors Hans-Peter Martin and Harald Schumann, in their 1997 best-selling book, *The Global Trap*, promulgated the notion of 'the 20:80 society' for the next century where '20% of the population will suffice to keep the world economy going'.[4] The predictions pendulum on the future of work has dramatically swung during the past twenty years from utopian to dystopian scenarios. Changes bring us both equality and inequality, and varied levels of access to new communications technologies—'high tech–high touch' for some, as John Naisbitt labelled it in *Megatrends*, but no access for others, or *high tech–no touch*. Writing in a special issue of *Futures*, Ziauddin Sardar and Jerome Ravetz

(1995) suggested that 'so deep is our collective ignorance of what cybertechnologies are doing to us that the first urgent task is to discover our ignorance'. What sort of revolution this is, where it is taking us, and what we can do about it, they lament, 'is as yet completely obscure'.[5] As a way of examining the major tensions, contradictions and issues to be addressed with contemporary communications, four paradoxes are discussed—the paradox of equity, the paradox of plenty, the paradox of users, and the paradox of diversity.

Paradox 1: The paradox of equity

The term 'globalisation' implies a commonality of world impact and effects. However, any careful examination of trends in communications towards globalisation must acknowledge that extraordinary disparities exist internationally between information rich and information poor. Despite the phenomenal growth of international communications in the last decade, serious disparities remain in terms of access to basic communications services throughout the world. Half of the world's population has no access to a telephone, and a third must travel two hours a day if they wish to use a telephone. A single country, Japan, has more telephones than another continent, Africa. High-income economies, with 15 per cent of the world's population, have 71 per cent of the world's telephone lines.[6]

Table 6.1 shows the serious disparity of access to major communication technologies, viz telephones, television and personal computers, in major world economies.

Table 6.1 International access to communications technologies, 1996 (main telephone lines, television sets and personal computers per 100 inhabitants in major economies)

Economy	Telephone density	TV density	PC density	Rank
United States	64.4	80.6	36.2	1
Canada	60.2	70.9	24.4	3
Australia	51.9	66.6	31.1	5
Japan	48.9	70.0	12.8	10
Singapore	51.3	36.1	21.7	12
Hong Kong	54.7	38.8	15.0	15
Israel	44.1	30.0	11.6	21
Russia	17.5	38.6	2.37	26
China	4.5	25.2	0.3	36
India	1.5	6.4	0.1	39

Source: *World Telecommunications Development Report: Universal Access*, International Telecommunications Union, Geneva, 1998.

These critical sources of global inequality should not merely be portrayed as critical differences at the macro level, as simply disparities between developed economies and developing or underdeveloped economies. There are also many deprived micro-cultures within the developed economies that do not have ready access to reasonably established communication technologies, let alone the newer technological systems such as the Internet. Many people in the ghettos of American inner cities are as seriously culturally deprived within rich America, as are those in the shanty towns of Africa, or people living in remote Australia. And, of course, these problems involve not only access to equipment, but related equity matters, concerning levels of literacy and technological awareness within different economies. Zealots of the electronic superhighway rarely discuss these vexed issues, and often forget that for many reasons there are large groups of people throughout the world who have not yet experienced 'the information revolution'.

As we build information networked societies, a series of vexed issues arise as to who has access to which services, on whose terms and at what cost. In privileged Australia, the issue of widespread access to the major communications services has traditionally been important, and many policy principles and practices have long addressed key issues of equity. As the 'old media' developed, they eventually provided generally high access rates. Most Australians have had good access to major communications facilities, such as a telephone and facsimile, radio and mainstream television, as well as to a national postal service and a local library. Urban dwellers have long been provided with more services than rural people. But despite the difficulty of our sprawling geography, particular attention has been paid to providing good rural access for major communication services.

Equity considerations have always been important in Australian communications policy which has a long history of debates about the implementation of social justice policies. Disputes over the role of the Postmaster-General's Department (PMG) at the time the Post and Telegraph Bill was debated in 1901 included equity issues:

> Should the department accept a policy of 'running at a loss' to assist the development of the country, offering services in distant and uneconomic places and making up the drain on departmental revenue through taxation? Or, should the department run its business as a 'commercial concern' balancing revenue and expenditure, and carrying out its programme through loans on which interest must be paid?[7]

The benchmark policy principle for much of Australian telecommunications during the twentieth century was *universal services at an affordable cost*. This was achieved in two ways—by state subsidies and by internal cross-subsidies. According to the Vernon Inquiry of 1974, telecommunications ran at a loss until the mid-1960s, which required the state to carry the losses of network building.[8] The extent of these losses is difficult to determine. They were clouded in consolidated accounts because, until 1975, telecommunications was part of the Australian Post Office. Clearly, though, the state-owned telecommunications carrier essentially reinvested in its own business, and never received huge payments from the Treasury to keep it afloat. It was actually internal cross-subsidies through the telephone tariff system that generated the bulk of the revenue required for network development. Telecom Australia long operated a complex tariff system where, for instance, business customers cross-subsidised residential customers, and urban dwellers cross-subsidised rural households. Hence this tariffing policy, which lacked the economic purity of charging particular customers according to direct servicing costs, provided substantial capital for infrastructure investment. The outcome was a national network accessible to almost all Australians.

This social principle was endorsed in legislation when Telecom Australia was constituted as a public statutory corporation in 1975. The Act stated that it must 'best meet the social, industrial, and commercial needs of the Australian people for telecommunications services and shall . . . make its telecommunications services available throughout Australia for all people who reasonably require those services' (section 6). Since then, telephone access rates have improved from 62 per cent of Australian households connected in 1975, to 80 per cent in 1982, to 95 per cent in 1990—one of the best performances of national household access to telephony in the world.[9] If economic rationalism had dominated public policy then, as it does now, Australia's leading-edge telecommunications network, generally agreed to be in the world's top ten national telecommunications networks, would never have been built.

However, with the new media we are now offering new or enhanced communications services on a user-pay principle. This means that people and institutions who have access to the new services are those who can pay directly for them. Mobile phones are extremely attractive, and the industry has grown like wildfire in recent years, yet only one in five Australians has a mobile telephone. Projections at the time of the introduction of subscription television in the mid-1990s suggested that by the year 2000 about 30 per cent of Australians would be on the system. The corollary, of course, is that 70 per cent of Australians will not then have access to these new television services.

The Internet may be wonderful, but how many in the community have ready access to it? Research based on Australian Bureau of Statistics (ABS) data of 1996 suggested that only 7 per cent of Australian homes were then fully digitally connected, though 25 per cent of Australians then had institutional access to the Net.[10] By August 1998, according to ABS data, Internet access had continued its remarkable growth to 18 per cent of all Australian households, although this is still fewer than one home in five.[11] So we appear to be building an information society with substantial disparities of access to the new media—a society of the information rich and information poor. The usual rejoinders to this are that 'everything in communications eventually becomes cheaper' and that access rates are likely to improve over time, or that 'we cannot all expect to own Rolls-Royces'. A critical issue about this era of new communications technologies and services is the extent to which the community will have widespread access to new services at affordable costs.

The new media further accentuate the trends of the widening gap between the haves and have-nots in Australian society. In 1972 the Henderson Commission of Inquiry into Poverty in Australia found that 12.5 per cent of households were living below the poverty line. Although the unemployment rate hit a low of 1.8 per cent in 1973, it has never been officially recorded as below 7 per cent throughout the 1990s. By 1996, according to the National Centre for Social and Economic Modelling, the poverty line had grown to almost 17 per cent nationwide, with a further 13.7 per cent classed as 'rather poor'. In summary, about 30 per cent of Australian households are now living in serious or marginal poverty.[12] How will these people afford access to new media?

Other changes in living modes and the excessive time demands placed on those in managerial or professional work have created a new polarisation in terms of interaction with new media. We are building a dichotomy between the 'time rich, cash poor'—those who have time to spend with media, but do not necessarily have the income to afford the subscriptions, as opposed to the 'time poor, cash rich'—people in high-income jobs, or dual-income households, who can afford the subscriptions but do not have the time to spend with media. (Students often comment that they belong to a 'time poor, cash poor' group![13]) The prime paradox of equity in our information society is that, at a time of boom times for the communications industry, the gap between rich and poor is actually widening, and we have constructed user-pay policies which further deepen the class divides as information rich/information poor.

As we move towards a more privatised, corporate, deregulated information economy, national principles of social equity have been dramatically challenged and no longer accepted as central policy objectives. Privatised telecommunications carriers of the 21st century will have shareholders' dividends as their prime responsibility. Will Australia's information society of the 21st century have universal services at an affordable cost as a benchmark policy principle across the full range of new information technologies?

Paradox 2: The paradox of plenty

Consumers of the world are now being bombarded with a range of communications channels and choices. Television services at first offered a few channels, but Australians can now pay to have 30 television channels if they wish, and some American cable companies offer their subscribers over 500 channels. Telephone companies have transformed their services from *plain old telephone services* (POTS), to offering more than 200 electronic communications services. The Japanese refer to these new services as *visual, intelligent and personal* (VIPs). We can direct dial to 230 countries using the telephone from our own homes. The flexible mobile phone has become extraordinarily popular as part of what is referred to as the move towards more 'personal communications'. The computer has moved from its warehouse location, to the desk, to the lap. We can shift text around at our convenience when word processing on our personal computer, copying, cutting and pasting for the better organisation of text. We can use the Internet to extract a remarkable range of information from around the world—a unique research tool for business, students and other individuals. All of these processes constitute a revolution for those who have access to these new modes of communication, for those who can afford the services and for those who are technologically communications literate.

Increased choice, however, does not necessarily transfer to satisfaction of needs. This is something that the old media appears to have in common with the new media. Unfortunately, there has been very little *qualitative* research done, by media organisations, advertisers, independent research bodies and academics, into the way in which people choose their media forms and how they relate to them. While there is a lot of survey work done on conventional television and radio ratings, and many programs and personalities in commercial media live or die according to the latest charts, the ratings show us virtually nothing about predispositions and preferences for programs. Rarely are we offered open-ended questions where viewers or listeners might be able to express their responses. Letters to the editor in daily newspapers, or listeners' letters to programs such as '60 Minutes', or talk-back radio, are marginalised attempts at meaningful audience feedback. One of the most important characteristics of traditional institutionalised media is the lack of meaningful participation for citizens and consumers. Audiences are essentially unknown and anonymous for most Australian media.

The new phase of 'plenty' in communications provides evidence of demand but not necessarily of satisfaction. The telecommunications industry has experienced what can accurately be described as exponential growth in the past two decades—Telstra has grown sixteen-fold in revenue terms during this period. In one of its best years, 1996/97, Telstra achieved a market take-up of more than a million digital mobile customers, its broadband network passed 2.1 million homes, it launched 65 new products and enhancements in one year, and Internet take-up grew by 15 per cent per month.[14] The growth of mobile phones has been phenomenal in recent years; Australia is in the world top ten for its fast take-up, and Telstra's mobile telecommunications services became a $2 billion business in 1997/98.[15] Telstra's chief competitor, Optus Communications, became a $2 billion revenue company within four years of its establishment in Australia, although its network roll-out costs and pay television problems took a heavy toll on its potential profitability. Meanwhile, as shown in Chapter 3, the established Packer and Fairfax media corporations are prospering from record growth in the past five years, and News Corporation has brilliantly gone global. About 500 Australian companies have emerged to try to find their niche in the dynamic digital world of multimedia. Collectively, the communications industry has been the best source of growth and opportunity in Australian industry during the past decade.

To what extent has this growth been translated to real consumer benefit? In telecommunications *the contemporary information revolution is primarily a supply-led revolution.* First priority is consistently given to building the new information networks, the high-capacity bandwidths to carry record volumes of voice, video and data as the superhighway delivery systems of the future. The leap of faith is that eventually the customers will somehow come to justify the unprecedented levels of network investment. Sandy Kyrish has suggested that communications history has many lessons of failed technology predictions from clearly aligned interest groups with 'biased' agendas. The predictions of the 1970s for US residential broadband networks are quoted as evidence, promoted by institutions and individuals drawing upon resonant themes of a utopian and revolutionised future to make optimistic and technology-driven predictions.[16] Howard Segal wrote of 'prophesy for profit's sake', citing the work of Alvin Toffler and John Naisbitt as emphasising positive, technology-driven futures because they satisfy corporate and mass markets.[17]

According to a *Communications Futures* report in 1994, the estimated cost of a hybrid optical fibre coaxial cable network, including customer premises equipment, to deliver services to all Australian households in the following decade (1995–2005) would be a staggering $41 billion.[18] This is essentially infrastructure cost, network extension investment by telecommunications carrier companies, and associated consumer spending on equipment, to bring into our homes comprehensive consumer services such as subscription television, Internet access, and on-line services such as home shopping, home banking, video-on-demand, home gambling and so on.

However, given the losses incurred by subscription television, the collapse of the subscription television company Australis, and the associated strategic reassessment of their infrastructure investments by Telstra and Optus, this figure became an unrealistically high estimate. In 1997 Telstra wrote down $342 million in the value of its pay television cable roll-out, and had another staggering $476 million write-off over pay television which was related to the company cutting short its cable roll-out from 4 million down to 2.5 million homes. Foxtel, the nation's largest pay television company, owned in association with News Corporation (and later PBL), lost $212 million during that financial year.[19] During fiscal year 1998, Telstra invested a further $92 million, and incurred losses of $83 million with Foxtel, as part owner.[20] In delivering the 1997 Charles Todd Memorial Lecture, the Managing Director of Telstra, Frank Blount, said:

> Communications belong to customers. Customers are now the driving force in the design and construction of the networks we are building: they are their reason and their touchstone . . . the era of build it and they will come is past: the time of come and we will build it is knocking at the door.[21]

Better sooner than later!

The point here is *not* to make a case against major investments in communication networks for the future, as indeed such is absolutely in the national interest, but to emphasise that *the demand side of the equation has been grossly neglected in strategic thinking about future communications services*. According to the Broadcasting Services Expert Group (BSEG),

> [A] broad range of networked services and applications projects is needed to develop Australia's technology base and demonstrate prototype

applications. There are opportunities in health, education, electronic commerce, transport, publishing, government services and many others.[22]

Just as Australia built a highly accessible world-class national digital telephone network during recent past decades, we now have to re-invest in upgraded higher-capacity networks to carry what is referred to collectively as the new on-line services. However, this must be accompanied by much more systematic investigation into estimating future demand for a range of new information services, and examining the associated social change. Questions that need to be answered include: who will be the users, what do they want or need in terms of information products and services, and how can we make the new services widely accessible and affordable?

There are so many behavioural issues to be resolved in relation to the new information services. How do we know, for instance, whether substantial numbers of people will want to shop from home via electronic shopping? How many gamblers will want to transfer to nambling (Internet gambling)? Will there be a major shift towards banking from home by phone, or will we still want the established banking interpersonal services? Will the banks forever maintain interpersonal banking at branches, rural and urban, to provide alternatives to technology-based banking in the future? Will we have any say? How many more companies will encourage or insist that certain members of their workforce be outsourced as teleworkers to save office space, power and lighting, and relatively expensive forms of superannuation payments, while they are company employees? There is a sense of a radical social change associated with the electronic superhighway. *What is needed is much more systematic research—corporate, policy, and academic social science-based—into the future communication needs of Australian consumers and Australian citizens.* This book will later further explore the complex trade-offs between new network investments and the evolution of a new economy built around the notion of an emerging on-line economy.

Further, what does the emergence of a plethora of new media mean for the 'old' media? The established media systems of print, radio and television are being surrounded and challenged by many different forms of new media. There are some dire predictions of the downfall of the old regimes of the press, publishing houses, commercial television and telephone companies, all allegedly threatened by this march of technological progress. George Gilder, in *Life After Television*, predicts the 'fall' of telephone companies and television networks as we have known them, and believes that the entire television culture will give way to what he calls 'teleputer culture', one of the more creative buzz terms that convergence has spawned. Just as the 1980s saw the collapse of 'the centralised scheme of a few thousand IBM mainframe computers and millions of dumb terminals', according to Gilder, we will see the collapse in the 1990s of 'the telephone scheme of a few thousand local central offices serving millions of dumb phones'.[23] Radical institutional change is on the way, if Gilder is right.

Predictions such as these defy the history of twentieth-century communications. Marshall McLuhan, in his classic work, *Understanding Media*, first published in 1964, predicted the death of print media. He was only outsold in the 1960s by the Bible—in print! The data on international and Australian sales in book publishing still continue to show consistently remarkable annual overall growth. The data in Chapter 3 also show that the established quality newspaper empire, Fairfax, remains a financial powerhouse based on the classified advertisements revenue generated by its major daily morning newspapers in Melbourne and Sydney. PBL, too, has enjoyed consistent growth in advertising revenue in the past decade, with prime time television remaining as its core business. The fact is that each major media form—print, radio, television—has actually shown remarkable adaptability in the face of new forms of media competition. Radio, for instance, was seen to be threatened by the introduction of television in the 1950s, yet it has evolved into something that enriches the lives of many Australians today as an intelligent discussion companion, and provides a richer musical diet than ever before. Australia has set world records for the take-up of new communications technologies, notably with the introduction of colour television, video cassette recorders (VCRs) and mobile telephones. For the period 1984–94, the VCR take-up soared from 26 per cent to 82 per cent of Australian homes, providing most homes with the technological capability to see many movies, yet the total cinema admissions rocketed from 29 million to 63 million during the same period.[24] As *The Economist* said:

> New technologies always contain with them both threats and opportunities. They have the potential both to make the companies in the business a great deal richer, and to sweep them away. Old companies always fear new technology. Hollywood was hostile to television, television terrified by the VCR . . . New technologies may not threaten their lives, but they usually change their role.[25]

New media have long tended to complement rather than replace old media. In the United States, for instance, network television vested interests fought bitterly against the development of cable television for decades in the 1950s and 1960s, but when the take-up of cable television reached more than 60 per cent of homes in the 1980s, the revenues and balance sheets of the established commercial networks showed they were still exceptionally strong businesses. Cable television never succeeded in knocking commercial network television out of the marketplace. In 1997, according to Paul Kagan Associates, America's big broadcast networks had revenues of US$12.7 billion, with profits before tax of US$697 million, whereas the cable networks had revenues of US$9.4 billion but gross profits of US$2.5 billion.[26] The 'old' commercial television networks still remained bigger in overall revenue terms, although the well-established cable networks had become the most profitable part of America's entertainment business.

One of the biggest future communications agendas is whether convergence will bring knock-out competition to the established media forms, resulting in the death of major media as is so commonly predicted in new wave magazines such as *Wired*. A more likely outcome is that we will see their continuity, complementarity and adaptability, just as has been the history of print, radio, television and telephony to date. Paradoxically, the old media seem to be thriving and prospering in this age of new media.

Paradox 3: The paradox of users

Technological abundance brings many dilemmas for the end users. We are awash with information in the advanced economies. Yet it does not necessarily follow that greater abundance of information automatically leads to better-informed users, or to a greater capacity to solve problems. So often, those selling the information superhighway suggest that our ability to solve problems, at so many levels, has been dramatically improved because of the new abundance of information. Yet is there any real sense that we are better now at finding solutions to problems, such as the drugs epidemic, road accident carnage, global pollution, the greenhouse effect, or the Irish or Middle East conflicts, because we have access to more information about these issues and know more about them? The chances should be greater, of course, and it is desirable to have as much information as possible to apply to problem-solving, but there are no guarantees that better decisions will be made. American academic Neil Postman has long challenged the commonly presumed nexus between amassing information and effective problem-solving, arguing that 'we are a culture consuming itself with information . . . information appears indiscriminately, directed at no one in particular, in enormous volumes at high speeds, and disconnected from theory, meaning or purpose'.[27] The challenge of the paradox of users is to match technological information abundance with meaning and purpose.

Nor does *more* necessarily mean *different* in the context of abundance of information. The 500 cable television systems in the United States, for instance, do not necessarily provide a greater diversity of programming than a conventional six-channel over-the-air Australian television model. Unless the ownership and control of the media outlets is ideologically diversified, the program diet is hardly likely to offer any real diversity. Take, for instance, the six-channel television model for an Australian city such as Melbourne; three channels (7, 9 and 10) are commercial networks, two channels are public broadcasters (2 and 28), and there is a community channel (31). It is remarkably diversified, in its own way, given its spectrum space limitation to six channels. To expand to a 500 cable television channel, as is possible now in many American cable systems, does not necessarily mean that there will be a more richly diversified programming choice. As the Bruce Springsteen song goes, 'fifty-seven channels and nothing on'. It is factors such as the nature of the ownership, the ideological mix in the system, the relative costs between imported and local production, managerial program judgments and the degree of user access that determine the diversity of content offered to the public.

So often, communication processes fail. It is astonishing how we are bombarded daily with a range of new information products and services, with such limited apparent consideration given to how consumers use, and interact with, the information product or service. A provocative article entitled 'I can't work this ?#!!@/ thing', which appeared in *International Business Week* in 1991, suggested that 'every day, across America, millions of managers, bankers, doctors, teachers, chief executives, and otherwise highly competent men and women are driven to helpless frustration by the products around them. The great revolution in electronic products that promised so much—speed, efficiency, and yes, fun—is not delivering. Office productivity isn't going up.'[28]

It is remarkable how many products of our information society fail with users because no consideration has been given to human factors. The above article, for example, cites Xerox's 8200 office copier which was packed with loads of on-board computer intelligence for collating, enlarging and reducing copy. This new product worked perfectly in the laboratory, but droves of loyal Xerox users could not make it work for them. Similarly, one of Japan's leading manufacturers of office equipment, Ricoh, found in a survey of its customers that 95 per cent of them never used the three key features it deliberately built into their machines to make them more appealing. Manufacturers, of course, would argue that they cannot do much about consumer laziness or ignorance, but they have tended to give all too little attention in their product design to how users will interact with their product. This is at the heart of the user paradox—an issue rarely discussed or researched in our information society.

David Sless, of the Communications Research Institute of Australia, has argued that whereas *good communications is about the transmission of signals from point to point, good communication (without the 's') is about the effective creation of understanding.* For Sless, the cumulative research evidence suggests that if you try to solve the problems of transmission without addressing the problems of understanding, there are few benefits, financial or social, to be had from the new information technologies. He invites people to apply a simple test in their office. 'Take the phone in front of you,' he says.

> How many of its features do you use? Do you know or understand what all its features are? Can you understand the instructions in the manual? Have you even looked in the manual? Now ask the same questions of the people around you. Unless your organisation is highly atypical, you will find that most people neither understand nor efficiently use even this most basic information technology.[29]

David Sless has suggested that the real IT issues to be explored are about how information technologies interact with people. He concluded from his applied research into these modes of interaction with new communications technologies:

> [T]he failures and cost blow outs occurred at the many interfaces between people and technology. We discovered that poor screen design was the norm; software was difficult to learn, documentation for all kinds of information technology was below unacceptable standard; and inputs forms were highly error prone and poorly designed. Add all these communication interface problems together and you will easily see why many of the potential benefits of the new technologies have been lost.

For David Sless, the communications marketplace has been rapidly filling with products that are indistinguishable from each other, and the ultimate difference between them will not be in the communications, but the communication. The information revolution has seen an extraordinary capital investment in equipment, systems and software. But do we know that productivity has really increased? He calls for a more constructionist view of communication, concerned with finding ways of mutually constructing our social realities. Hence, in information design in the production of web sites, he advocates 'a shift from the production of Web sites per se to the management of the relationship between an organisation and their publics—communication which meets the needs of both the client and the user'.[30]

There has been too little user-based communications research, and the application of this kind of thinking has been limited. American researcher Brenda Dervin has encouraged people to broaden the way they used the term 'information'. She suggested that until the mid-1970s 'information' was closely related to decision-making—information was too often seen as a matter of transmission rather than a process of construction around users. Dervin broadened the notion of 'information' to encompass understanding, clarification and 'personal sense'. For Dervin, the real complexities are around seeing information as informing, and as 'sense making, meaning making and communicating'.[31] If this is accepted, then the focus must shift from information systems and what they can provide—the supply side—towards users, and how they make sense of new communications technologies—the needs and demand side.

One interesting manifestation of the dominance of the supply side of the process is seen in the nature of much of the language employed in communications services. In the computer and telecommunications industry, the language used to describe consumers is consistently about 'penetration'—such as a 'penetration' rate of 30 per cent for pay television, or a 60 per cent 'penetration' rate for mobile phones, rather than using terms such as 'access' or 'take-up rates' by customers. We tend to talk about 'customers' rather than 'citizens', because 'customers' implies they can pay. This is more than semantics, because so much of the language suggests an essentially one-way uncommunicative process based on consumption. Indeed, the term 'penetration' suggests a highly aggressive one-way act. Similarly, the term 'hits' is used to refer to the number of Internet users who access a particular site. And, of course, we sit at a 'terminal' to access the Internet. One would assume that interactivity is supposed to be a kinder, more communicative process!

Paradoxically, we offer more services than ever before, yet we seem to understand little about the people who use the communication processes. Major media have traditionally offered their audiences virtually no open-ended qualitative process of investigation about audience response to programs. The management of telecommunications carriers has been dominated by engineers who were superb at building networks. Only in recent times has the notion of 'delighting customers' emerged, but suppliers have embarked upon only limited systematic social science-based research into customers' needs. And in information technology, the first law of computing was that 'applications expand so as to fill the bandwidth available'. If the managers of 'new media' show as little real curiosity about finding out about their audiences and customers as those of the old regime, they may be slaughtered commercially in the new highly competitive regime.

A major paradox of the communications process is that it involves such limited communication in the sense of effective creation of understanding. Michael Patkin, a past President of the Ergonomics Society of Australia, has suggested that 'we need to stop thinking about the information society and start thinking about the user society'.[32] *The real issues to be addressed about the future of our information society are not what the technologies are going to be like but what* we *are going to be like.*

Paradox 4: The paradox of diversity

The power relationships of major media and communications institutions are changing in a globalised society. We have entered a major phase of more globalised media corporations in a world with more strongly commercialised, more globally oligopolistic media systems. With the emerging new media come higher-capacity network platforms, new forms of distribution and the increasing global reach of new media forms. Paradoxically, at a time of institutional consolidation through more corporate mergers and alliances, audiences are now more fragmented through globalisation and more culturally diversified.

Media ownership patterns are not only changing *within* nations; global forces are also again at work. Just as the ownership and control of the world's oil and automotive businesses became progressively more concentrated during the twentieth century, many media businesses are now becoming part of a global entertainment oligopoly. The formation of alliances as a product of convergence has meant that many former great media companies have now been subsumed into larger entertainment conglomerates. As *The Economist* argues:

> These days the grand old names of entertainment have more resonance than power. Paramount is part of Viacom, a cable company; Universal part of Seagram, a drinks and entertainment company; MGM, once the roaring lion of Hollywood, has been reduced to a whisker because it is not part of one of the giants. And RCA, once the most important broadcasting company in the world, is now a record label belonging to Bertelsmann, a German entertainment behemoth.[33]

Seven huge corporations—Time Warner, Walt Disney, Bertelsmann, Viacom, News Corporation, Seagram and Sony—dominate the world's publishing and entertainment business.

In broadcasting, several key technological and economic factors are driving the greater global spread of media. First, better communication distribution platforms, such as satellite and cable delivery systems, have enabled vastly greater global geographical reach and distribution. There is a *greater spread of service delivery platforms*, notably satellite-delivered television beamed to multiple countries, such as News Corporation's Hong Kong-based Star TV broadcasting to 53 countries in Asia with a total population of 2.5 billion people and an audience of 220 million viewers. Second, there has been the *emergence of global television networks*, such as Turner's CNN for news, Home Box Office (HBO) for movies, and ESPN for sports. Ted Turner first conceived the idea of a 24-hour television national news network, CNN, in 1980 and it became the most successful channel on American domestic cable television. CNN is now the world's most ubiquitous international news network, broadcasting to over 200 countries. The BBC, one of the world's best public service broadcasting organisations, has been the most successful public broadcaster to internationalise its program distribution. Third, *globalisation of distribution has spawned greater co-production alliances between the major media corporations and local production houses*. In theory, economies of scale that come with globalisation ought to provide the significant commercial benefit of larger markets. The Murdochs and Turners can broadcast to dozens of countries, with potentially lucrative advertising rates that might be charged in such big markets to the McDonald's, Toyotas and Nikes of this world. In fact, though, this has not actually happened, and localisation of advertising and programming have remained important to attracting audiences. As the *Asian Communications Handbook* 1998 points out:

> Realising the need to provide programs in Asian languages, the multinational satellite companies, such as Star TV and the BBC, have launched a series of television productions in association with Asian companies. For instance, the BBC has commissioned a series of productions in India; CNN has established its bureaus in Asia; and Asian companies in Hongkong, Taiwan, Singapore and India have also commenced television production aimed at regional audiences in Asia. Local television stations have launched new programs to win back the audiences lost to Star TV.[34]

Although we have seen the emergence of a more globalised economy based on global information networks in the past decade or so, have we seen the emergence of more globalised cultural products? Media forms are cultural artefacts which are inevitably richly diversified in content and form. While we may now be able to deliver television programs around the world through extraordinarily sophisticated international delivery networks, has this produced global television in cultural terms? CNN, for instance, broadcasts in more than 200 countries, and although its programs are globally distributed, surely it does not offer 'global news' in cultural terms.

The trend towards greater globalisation of media systems raises profound questions about its impact on identity. A person's identity is constructed from their experiences and from the sources of meaning in their lives. Individuals and nations have a plurality of identities. The key institutions in society—state, church, school, university, union, political parties, community associations—are important influences in the social construction of our identity—of what we are, or think we are. Media and communications institutions are central players in influencing and legitimising individual and national identity. What we choose to see as critical issues in society depends in part on how the media constructs its agendas of discourse. It is not that the major media institutions of this century—press, radio, television—have manipulated or dictated to audiences *what they must think*; rather, they have influenced *what they can think about*. Ultimately, our responses to constructed agendas by media are highly individualistic.

Anthony Smith has pointed out that our new information systems are able to bring into the living rooms of everyone who possesses a television set, images and experiences from every part of the globe. He invites the critical questions:

> [I]s this power helping to overcome national divisions or does it serve to reinforce them? Are the mass media simply overlaying ethnic and national cultures with a cosmopolitan veneer, or are they actually producing a new global culture that will ultimately replace the older national cultures that divide humanity today? Is television the producer, or merely the medium, of this global culture? And, what in any case, would be the content of this global culture?[35]

Although there are some key trends in terms of global programming and distribution, the issue of whether a culturally integrated global television market is emerging is hotly disputed. Clearly, there is now greater international reach of international events. So many more people throughout the world are now able to experience major events on television. Take, for instance, the wide international television coverage given to events such as the Gulf War, the downfall of Presidents Marcos of the Philippines and Sukarno of Indonesia, the death of Princess Diana and the impeachment trial of US President Bill Clinton. With the broadcast of international sport we have come the closest of all to the realisation of McLuhan's 'global village', with global audiences of more than 100 million in some cases watching the Olympic Games, the

World Cup in soccer and cricket, the Grand Slams of tennis, including Wimbledon and the Australian Open, America's Superbowl, and multiple Grand Prix in motor racing. There are now television audiences of unprecedented size for particular programs, especially in sport.

National regulation can, of course, restrict the spread of imported programs and the introduction of technological systems which deliver them, and many governments have grappled with the complex issue of whether they should allow new distribution systems to openly compete with established domestic media, or whether they should try to stop the tide of technological change. Several Asian governments in the past decade, such as China, have banned the entry of satellite dishes which bring foreign programming into their country, only to find that such legislation cannot be adequately policed.

Inevitably there are tensions between the commercial interests of national broadcasters and local regional responses and sensitivities, as Rupert Murdoch has found. Murdoch argued in a major speech in 1993 when attempting to sell 'the communications revolution' that new media technology posed 'an unambiguous threat to totalitarian regimes everywhere'.[36] This was highly offensive to the Chinese government and threatened Murdoch's future access to their huge television market. The Chinese government had also objected to the BBC's version of events in Beijing's Tiananmen Square in 1989, and to a BBC documentary which had portrayed Chairman Mao Zedong as a shocking tyrant. Murdoch subsequently removed the BBC World Television Service from his Asian satellite service. Murdoch was later alleged to have repudiated an agreement to publish a controversial book written by the last Governor of Hong Kong, Chris Patten, who had infuriated the Beijing government during the last days of colonial rule before the resumption of sovereignty by China on 1 July 1997. Similarly, in 1995, a Star TV program beamed into India created great offence locally. The chief guest on the program referred to the idolised Mahatma Gandhi as a 'Baniya bastard'. (He meant by this that Gandhi was a miser.) Indian parliamentarians were outraged by this and responded by accusing Star TV of cultural terrorism. A warrant was issued under the Indian law of criminal liability for the arrest of Rupert Murdoch!

More widely than these well-publicised incidents, the increasing trend towards the globalisation of television has led to an old set of debates and concerns being revisited in new contexts—the issues referred to as *cultural imperialism*. The central proposition of cultural imperialism is that certain dominant cultures threaten to overwhelm more vulnerable ones (for example, America over Europe or Asia, and Western domination over the rest of the world). In 1971, Ariel Dorfman and Armand Mattleart's *How to Read Donald Duck: Imperialistic Ideology in the Disney Comic* argued that Disney comics denigrated Third World cultures and contained covert ideological messages that American superiority was in everyone's best interests. American academic Herbert Schiller was a prolific writer over several decades about the alleged undesirable effects of the domination of the American audio-visual industry over other cultures. His attacks were focused on transnational corporations (TNCs):

> They function as private profit making enterprises seeking markets, which they term audience. They provide in their imagery and messagery the beliefs and perspectives that create and reinforce their audiences' attachment to the way things are in the [capitalist] system overall.[37]

The media imperialism thesis has been challenged on many grounds. First, it is too simplistic to argue a cause-and-effect relationship, with an active First World simply forcing its products on a passive Third World. The work of Ien Ang, a cultural studies theorist, raises important questions about regarding soap operas as global forms of popular culture. Her research on the soap opera 'Dallas' argued that viewers interpret the program in a different cultural context from its source, and that their reading of the text of 'Dallas' is influenced more by context than by origin. Hence, it makes little sense to see the overseas production houses of the world as agents of cultural imperialism. As Brian Shoesmith has suggested, '[S]top seeing the introduction of the new communication technologies in terms of technology transfer, which Pacey correctly argues always implies an active donor and a passive recipient, and begin to see it in terms of technology dialogue, which implies an entirely different set of power relationships.'[38]

Second, global 'mass culture' does not so much replace local culture as co-exist with it. Research, for example, into the content analysis of television programs in nine Asian countries showed:

- In seven of the nine countries, more hours of domestic television programs were broadcast than of imported programs (the exceptions were Malaysia and Singapore).
- The four countries with the greatest proportion of American imports have relatively large English-speaking populations.
- The share of imported programs in total national local television supply was often low—for example, Korea 8.9 per cent, India 8.3 per cent, Philippines 32.7 per cent (cf Australia 54.2 per cent).[39]

Third, the imported 'mass culture' can be indigenised, put to local use and given a local accent. As a senior management executive of Star

TV explained, '[T]he first decision we made was that broadcasting one stream of programming to 53 countries did not make sense.' Going local meant 'observing the local language, recognising the local culture, local sensitivities and local market conditions. Also, devolution of more management responsibilities to regional activities . . .'[40] Hence, it makes sense in commercial terms to form such cultural alliances for production.

Fourth, there are powerful reverse currents, as a number of Third World countries (Mexico, Brazil, India and Egypt) in fact dominate their own markets and have become exporters.[41] This trend is even more strongly developed with cinema, as Ella Shohat and Robert Stam point out:

> [T]he cinemas of Africa, Asia, and Latin America clearly constitute the majority cinema in the world. Third World cinema, taken in a broad sense, far from being a marginal appendage to First World cinema, actually produces most of the world's feature films. If one excludes films made for TV, India is the leading producer of fiction films in the world, producing between 700–1000 feature films a year. Asian countries, taken together, produce over half of the world production.[42]

For global film production the evidence shows falling Hollywood domination: Edward Herman and Robert McChesney point out that although film output 'is dominated by studios owned by Disney, Time Warner, Viacom, Universal Sony, Polygram, MGM, and News Corporation . . . the non-US revenue increased from 33% in 1984 to 50% in 1993, a continuing trend'.[43]

The cultural forces impacting upon our lives, and fashioning our identity, have a great potential to offer us richer and much more diversified media as a result of the information revolution. A key paradox of the new media is institutional consolidation on the one hand, yet user diversification on the other. Corporations search for economies of scale as a result of globalisation, yet find that their markets are more fragmented and complex culturally than ever before. Diversity is alive and well in the world's many global villages.

7 Towards an information society

Knowledge is critical for development. If we want to live better tomorrow than today, if we want to raise our living standards . . . we must do more than simply transform more resources, for resources are scarce. We must use those resources in ways that generate ever higher returns on our efforts and investments.

Knowledge for Development, World Bank, Washington, DC, 1998/99, p. 16

The group believes that, rather than seeing the communication network as a system that connects us to phones, televisions and computers, we should see it as a platform underpinning our society, supporting a diverse and interwoven range of social, business and community activity.

Networking Australia's Future, Report of the Broadband Services Expert Group (BSEG), 1994, p. 3

Dominant forms of communications technologies have shaped the character of civilisation at different points in history. Canadian academic Harold Innes, who investigated the different roles that clay tablets, papyrus, parchment and paper had in shaping different civilisations, believed there was a fundamental difference between speech-oriented and literate cultures.[1] Print culture emerged from the portable and preservable nature of paper, which together with the invention of the printing press, facilitated widespread literacy and enabled the creation of our modern legal and political institutions as pillars of the nation state. American academic Ithiel de Sola Pool argued that the printing press brought about modern democracy.[2] During the twentieth century new forms of electronic communications have progressively challenged the primacy of print. In the 1920s, radio demonstrated the dramatic impact of the audio immediacy of its mental pictures, and some decades later television in the

1950s and 1960s emerged as the dominant medium of mood and impression, saturating our consciousness with mosaic images as advertisers' dreams came true. Computers gave us data as bits—abstractions in a shift towards more symbolic forms of communication.

Twenty-five years ago, Daniel Bell, Professor of Sociology at Harvard University, coined the term 'post-industrial society' to describe the emergence of a new social framework where economic activity and social exchanges would centre on the way knowledge is created and retrieved in ways that would determine the character of our occupations and work. He argued that with the shortening of labour time and the diminution of the production worker, knowledge and its applications had replaced labour as the source of 'value added' in the national product. Writing in 1973, he argued that the United States had changed from a 'goods producing to a service society' where the net new growth in the previous two decades had been entirely in the area of post-industrial services. For Bell, 'in that sense, just as capital and labour have been the central variables of industrial society, so information and knowledge are the crucial variables of post industrial society'.[3]

Bell's notion of an emerging knowledge/service-based economy has now come to international fruition. In an era of multiple societal changes, the information revolution is widely regarded as the most far-reaching collective shift of our era—an epochal change. Never before have so many people been able to generate, store, transmit and retrieve so much information in so many different forms. So many new communications technologies and services have appeared in recent years—satellites and fibre optics to carry telephony and data, video cassette recorders, pay or subscription television, multimedia and on-line services, intelligent and interactive networks as electronic superhighways, the Internet with its World Wide Web, and so on. Collectively these systems and services point to a new era of communications for many societies. Contemporary advanced society has been relabelled an information society, or the intelligent state (Connors, 1993), or knowledge society (Drucker, 1994), the third wave civilisation (Toffler, 1995), the infomedia age (Koelsch, 1995), or cyberspace (*Time*, Special Issue, 1995) and cybersociety (Jones, 1995) or the network society (Castells, 1996).[4]

Such advocates have different perceptions of change, but they share the claim that advanced societies are being transformed by a revolution in information technology. New kinds of industrial societies have emerged in recent history from new sources of energy, such as fossil fuels, or electricity or nuclear power. The information society school of thought argues that for the past two decades or so we have been progressively undergoing structural economic change, in that traditional manufacturing industries no longer remain the core of national wealth and employment generation. Nowadays, a range of activities associated with information-based goods and services are commonly seen as the prime creators of national prosperity. In recent years, many forms of communication industries and practices have experienced unprecedented growth to become the prime catalyst for investment, industrial development, market share and employment. A new economy is seen to have emerged around the industrial 'newcomers' such as media, multimedia and entertainment corporations, telecommunications carriers and service providers, information technology companies and the Internet industry with its associated on-line services. We may be experiencing the reformulation of growth capitalism around these goods, services and practices, with the economic dynamism of the United States during the 1990s regarded as the prototype of the new capitalist information economy.

Attempts to put together a global picture of the size of the information economy show huge variations in the estimates. According to Arthur Andersen's *1998 Report on the Communications, Media and Entertainment Industries*, the world's converging communications industry had become a US$1.1 trillion juggernaut by 1998, expanding by about 10 per cent annually since 1993. As a benchmark for this figure, a revenue of US$1 trillion is more than the individual GDPs of 139 countries, including Argentina, Australia, Canada, Belgium and Korea. Some 150 of the world's leading communications firms earned US$776 billion of that US$1.1 trillion. Since 1995, the report argues, more money has been spent on telecommunications services alone than on oil, representing a new kind of energy shift.[5] In the United States, the information technology sector now constitutes 8.2 per cent of total GDP, and is growing at twice the rate of the economy as a whole.

Meanwhile a comparable survey by the International Data Corporation argues that communications was responsible for US$1.8 trillion in spending internationally in 1997, or approximately 6.5 per cent of aggregate global GDP: spending on communications in 1997 was nearly 40 per cent greater than in 1992, growing 27 per cent faster than the overall worldwide GDP which had grown by an average of 5.5 per cent annually during those years.[6] An OECD survey argued that the world IT market had a compound growth rate of 9.9 per cent for the period 1987–94, which compared with a world growth rate of 5.7 per cent for the same period.[7] Symbolically, Microsoft replaced General Electric at the top of the market capitalisation tables in late 1998 in the United States. These data are indicative of profound economic structural change in the last quarter of the second century of world capitalism.

At one level, these major variations in economic estimates of the information economy indicate methodological variations, particularly boundary differences for the various sectors which make up the collective area of communications. The Arthur Andersen report centred on media and entertainment services, while the IDC report focused on information technology and telecommunications, to produce large variations. Linking such substantial growth of expenditure on information-based goods and services to gains in real productivity is also extremely difficult. How can anyone prove a nexus between the huge investments in computerisation and real gains in corporate or national efficiency? While the level of growth and its real impact is problematic, the trend line is clear that major investment in communications goods and services is now vital to national prosperity.

International communications development policies

Governments of nation states are attempting to reposition their economies within the new interdependent global information economy so as to increase their relative share of wealth and power. The leading economies of Western democracies have embarked upon growth strategies with a critical objective to build an efficient and dynamic domestic convergent communications industry that can find its place within the arena of international competition. Virtually every developed economy has a national plan, or set of strategies and policies, seeking their own comparative advantage for the development of their communications industry, and the associated benefits that come with these enabling technologies.

Paradoxically, at a time when virtually every government in developed countries is embarking upon greater privatisation, deregulation and liberalisation of their communications industry, and withdrawal from many of their traditional roles, most governments have been constructing national communications development strategy plans. Their common assumption is that no future economy will grow substantially, or attract foreign investment and technology, unless it has a world-class communications infrastructure. Visionary blueprints are also seen as good promotional tools for domestic politics. These national development plans tend to be highly technologically deterministic, centred on the urgent development of high-capacity information infrastructure, and generally with only sketchy attempts at cost-benefit analysis or analysis of its social impact. Examination of the issues about supply rather than demand invariably get first priority in national communications development blueprints. The services that might be offered, or social considerations about users' information and future services needs, are generally only sketchily considered as part of the process of their strategic thinking. There has often been an extraordinary oversell of attempts to draw up national communication development policy blueprints, where the electronic superhighway of the future has been marketed in such a way that it is the 'superhypeway'. There has almost been an undeclared international competition for who can construct the most ambitious vision of their nation's blueprint for their cyberfuture!

The extent to which there is any relationship between centralised national planning for communications policy of the future, the clash of domestic vested interests, and what is really being achieved in terms of following the blueprint, varies widely from nation to nation. Since the mid-1990s, however, there has been no absence of communications development policy documents emanating from the governments of developed nations. These highlights may be briefly summarised.

United States

In the mid-1990s the US government outlined a plan for a National Information Infrastructure (NII) to transform its domestic economy based on broadband networks. The international extension of the NII envisaged by the United States is the Global Information Infrastructure (GII). Speaking at the First World Telecommunications Development Conference in 1994, US Vice President Al Gore said, '[W]e now have at hand the technological breakthroughs and economic means to bring all the communities of the world together.'[8] A close reading of the plan indicates more ethnocentric motives from the world's 'dominant culture'. Gore's vision referred to:

> . . . a planetary information network that transmits messages and images with the speed of light from the largest city to the smallest village on every continent . . . these highways will allow us to share information, to connect, and to communicate as a global community . . . it will consist of hundreds of different networks, run by different companies and using different technologies, all connected together in a giant network of networks, providing telephone and interactive digital video to almost every American.[9]

Gore was not aware in 1994 that the Internet was about to grow so fast that it would become 'the planetary information network' of this vision. Glowing pronouncements were claimed for the national benefits by the authors of the National Information Infrastructure Progress Report of September 1994, such as ' . . . to reduce health care costs by some $36 billion per year, prepare our children for the knowledge based economy of the 21st century, add more than $100 billion to our Gross Domestic Product over the next decade and add 500,000 jobs by 1996 whilst extending the quality of work life'.[10] The GII was intended to facilitate economic progress, enhance democratic trends and provide better solutions to global environmental challenges.[11] The plan is based on five principles: encouragement of private investment, promotion of competition, creation of a regulatory framework that can keep pace with rapid technological and market changes, provision of open competition to the network for all information providers, and ensuring the provision of universal service. These five principles now consistently appear in so many economic development national blueprints that they have virtually become international communications policy benchmarks. However, although the US government is adept at issuing visionary documents, it would be quite misleading to suggest that their information economy is an outcome of state architecture. There is no country in the world so fiercely committed to individual enterprise and private sector achievement and reward as the United States, especially in communications.

Europe

In European planning, too, the development of a superhighway was seen to encompass a global infrastructure. In December 1993 the European Council requested a report on information infrastructures. Subsequently, the Bangemann report (1994) argued that the first countries to enter the information society 'will reap the greatest rewards' and it suggested little choice, otherwise Europe would face declining investment and a job squeeze. The Bangemann recommendations to the European Commission suggested that those countries which 'temporise or seek half hearted solutions' will face disastrous consequences. The only question, according to this report, was whether this will be a 'strategic creation for the whole Union, or a more fragmented and much less effective amalgam of individual initiatives by individual states'. The Bangemann report proposed ten applications to launch the information society which 'have the potential to improve the quality of life of Europe's citizens, the efficiency of our social and economic organisation and to reinforce cohesion'.[12] The best research and thinking about the information society, vexed policy issues and social factors has come from the European Commission, the United Kingdom and Canada in the past few years.

Asia and South-East Asia

Japan's Telecommunications Council of the Ministry of Posts and Telecommunications 1997 report, *Vision 21 for Info-communications*, argued that 'info-communications' are an economic tool for change that will rebuild 'economic and social factors in new forms' to revitalise the Japanese economy. The plan postulated a massive investment in Japan's communications networks, at a projected capital investment of a staggering 7.2 trillion yen by 2010, intended to create about 2.44 million new jobs.[13] The recommended plan was to build total digital networks that would seamlessly interconnect cable/radio and mobile/fixed networks. The plan was to 'complete the laying of subscriber fiber-optic networks around the country by 2010 and develop and deploy subscriber wireless access systems: for mobile networks, develop and deploy by 2000 next generation mobile phone systems that meet world standards, develop and deploy multi media mobile access systems, and develop and deploy by 2010 the next generation LEO (satellite) system'.[14] Since this technologically deterministic visionary report was written in 1997, Japan's economy has moved into a serious recession.

In South-East Asia a group of information technology professionals captured government strategic policy in the early 1990s, and a set of extremely ambitious IT development plans emerged as top national priorities. Singapore's remarkable National Computer Board's 1992 *Vision of an Intelligent Island* (pre-US Vice President Al Gore's plans, and before America's NII and GII) presented a vision of a highly informatised society, developing Singapore as a global hub for services, transportation and business. The Singapore plan, driven steadfastly by its government, is also centred on the notion of an advanced nationwide information infrastructure:

> In our vision, some 15 years from now, Singapore, the Intelligent Island, will be among the first countries in the world with an advanced nationwide information infrastructure. It will interconnect computers in virtually every home, office, school, and factory. Text, sound, pictures, video, documents, designs and other forms of media can be transferred and shared through this broadband information infrastructure made up of fibre optic cables reaching to all homes and offices, and a pervasive wireless network working in tandem . . . A wide range of new infrastructural services, linking government, business and the people, will be created to take advantage of the new broadband and tetherless network technology.[15]

This small island state, with so few natural resources, has developed its own high-technology infrastructure that has brought with it considerable national wealth and a more stable economy than most of its neighbours. Singapore further advanced its high-capacity network blueprint in 1997 towards the Internet as the major communications platform with its *Singapore One: Beyond the Internet* (1997) in a plan intended to deliver 'a potentially unlimited range of multimedia services to the workplace, the home and the school'.[16] The government of Singapore, however, has long been much stronger at developing economic/high-technology initiatives than they have at fostering expressive cultural policy.

Singapore's neighbour, Malaysia, embarked upon a comparable plan in the mid-1990s, the Multimedia Super Corridor (MSC). This is promoted as being 'not just another technology', but as a 'multimedia utopia'. MSC is marketed as a 'gift from the Malaysian government to technology developers and users to expand their Asian presence'. Such was the level of commitment to this plan that for some time in mid-1997 the Malaysian Prime Minister, Dr Mahathir, stood aside from his day-to-day leadership duties to personally promote the MSC in North America and Europe. MSC involved the construction of a 15 × 50-kilometre 'supercorridor' south of the downtown capital city Kuala Lumpur, and a world-class physical infrastructure and 'next generation' 2.5–10 gigabit multimedia network. The government documentation suggests that Malaysia is creating the 'perfect environment' for companies wanting to create, distribute, and employ multimedia products and services. But this M$20 billion dream hit hard times during the Asian economic crisis. Investors became scared, and the project had to be scaled back significantly from its ambitious conception. In a nation not renowned for its openness, there are ambitious expectations of 'special cyberlaws, policies, and practices tailored to enable residents to achieve the full promise of multimedia'.[17]

More widely for Asia, the first APEC Ministerial Meeting on Telecommunications and Information Industry, held in Seoul in May 1995, advanced the concept of an Asia Pacific Information Infrastructure (APII). The primary goal was to establish a seamless APEC-wide information infrastructure, within a cooperative framework of APEC members, by interconnecting their individual national information infrastructures. Five objectives of, and ten principles for, the APII were agreed as a significant step 'towards the realisation of the 21st century information society'. The ten core principles for the APII were:

1. Encouraging member economies in the construction of domestic telecommunications and information infrastructure based on their own reality.
2. Promoting a competition-driven environment.

3. Encouraging business–private sector investment and participation.
4. Creating a flexible policy and regulatory framework.
5. Intensifying cooperation among member economies.
6. Narrowing the infrastructure gap between the advanced and developing economies.
7. Ensuring open and non-discriminatory access to public telecommunications networks for all information.
8. Ensuring universal provision of access to public telecommunications services for all information providers and users in accordance with domestic laws and regulations.
9. Promoting diversity of content, including cultural and linguistic diversity.
10. Ensuring the protection of intellectual property rights, privacy and data security.[18]

The most difficult question for the member economies of APII in the next few years is the extent to which such ambitious development plans are realisable in view of the serious economic problems that have emerged for many of those economies since late in 1997. However, the presumed nexus between future prosperity for developing countries and building an advanced information society has not changed, and there are indications in 1999 that communications is most likely to remain the top priority for strategic investment.

Australia

And what about Australia's position in terms of the construction of a national development blueprint for a convergent communications industry? We have had a plethora of reports, reviews, inquiries and working papers in this major policy field in recent years, but no consolidated national strategic plan. A compendium of Commonwealth government reviews and inquiries in communications policy, published in 1995, showed no less than 81 reviews/reports/inquiries in recent years on behalf of the national government into various aspects of communications policy. The Centre for International Research on Communication and Information Technologies (CIRCIT) showed that eleven different federal government departments were involved, incorporating the activities of some 79 government-constituted agencies and ad hoc working groups, and eight Parliamentary Committees.[19] These reviews include generic economic analysis, such as the Hilmer report into national competition policy, reviews into radically changing telecommunications policy, new broadcasting issues, market segment analysis, and specialised reviews into particularly vexed policy areas such as copyright and privacy problems. One response to such bureaucratic business is to suggest that this represents a national government using comprehensive investigative mechanisms to keep in touch with contemporary economic and social change; presumably everyone would be most critical of an Australian government that was not monitoring such momentous changes. This would, however, be a generous interpretation of what the nation gained from the communications inquiry machine, which some cynics argue is growing faster than the industry itself!

Nevertheless, there have been many valuable, but individualised, reports into important aspects of Australia's communications future. One of the most notable of the commissioned reports was *Commerce in Content* (1994) which focused on interactive multimedia markets and contributed to raising national consciousness about the growth of multimedia. The government issued *Creative Nation* (1994), an imaginative cultural policy paper which seemed quickly forgotten. *Communications Futures* (1995) focused on residential communication networks for the future, and raised important methodological issues relating to supply and demand, especially drawing attention to the staggering proposed levels of investment in infrastructure with so little attention to the future demand side. An ASTEC report, *Surf's Up: Alternative Futures for Full Service Network in Australia* (1995), probed issues of the take-up of new services and offered valuable predictive scenarios for new networked services to 2010. The Broadband Services Expert Group's report, released in 1995, analysed evolving broadband networks and services and urged a national strategy focused on harnessing the benefits of information and communication services development. Their prime call was that 'a central element of that strategy should be a managed evolutionary approach . . . Co-ordination between all participants will be vital . . . The strategy would draw together several key elements: education and community access; industry development; and the role of government. *The most important part of the strategy is leadership. We need leadership at the highest levels, within government and the community.*'[20]

Following the BSEG, the then Prime Minister, Paul Keating, announced the creation of a new National Information Services Council (NISC) in April 1995, with the task to construct a national strategy for information and communications services and technologies which recognised the important contributions of the BSEG, ASTEC, the *Communications Futures* report and others. The statement emphasised the importance of national policy development centred on government use of information technology, and information and communication technology industry development measures to encourage technological diffusion across economic, industry, educational and societal settings. NISC, the National Information Services Council, subsequently produced valuable discussion papers from four talented working parties. Again, NISC reiterated the importance of national leadership in advocating that 'Australia needs a national vision for information and communication technology and services industries, to provide a framework for co-operative activity within and between industry and governments, federal and state'.[21] Regrettably, NISC met only once as a full group, on 10 August 1995. A promising attempt to integrate communications policy faltered.

More recently, a series of reports commissioned by the Commonwealth government have dealt with the contentious policy area of industry development, and with the development of our on-line services. For the 1990s a series of professional inquiries, investigations and reports have produced a great deal of valuable data, raised awareness of key issues and proposed a whole series of major policy changes.

What has *not* emerged, however, is a coherent, credible and achievable Australian national communications economic and social development blueprint from any Australian political party, or any collective corporate interests, research institution or Australian government. We shall return to this major issue in the final chapter.

Australia's convergent communications economy

Technological and economic change has seen the emergence in the past decade of an Australian communications industry where formerly disparate sectors of media, information technology and telecommunications have come to converge, functionally and institutionally. We can now talk of an Australian convergent communications industry. In economic terms it now makes sense to look at wider sectors of the economy than previously, where a range of information goods and services contribute significantly to national economic activity. There is a need to analyse media economic data, for instance, that not only takes in traditional content forms, such as newspaper, radio and television revenues, but also includes newer media forms such as pay television, video rentals and computer games. Telecommunications revenue now incorporates not only its staple diet, plain old telephone services, but many forms of newer data services, as well as directories, traded information services and electronic commerce. Moreover, also included in economic data for a convergent industry is the chain of suppliers involved in the electronics industry, bringing in manufacturers who build telephone exchanges or mobile telephone base stations, as well as developers of computing software.

The performance of Telstra during the 1990s symbolises an important shift in the structure of the Australian economy. Telstra became our industrial market leader when it posted the greatest profit ever in the nation's corporate history (1997/98) and the first company in Australia to reach a market capitalisation in excess of $100 billion (1998). No longer

Table 7.1 Australia's communications industry: The top 30 companies, 1997/98

Company	Net annual revenue ($m)	Publishing date
Telstra Corporation Ltd	17 296.0	June 1998
Cable & Wireless Optus Pty Ltd	2 933.0	June 1998
IBM Australia Ltd	2 214.0	January 1998
The News Corporation Ltd	1 852.0	30 June 1998
Publishing and Broadcasting Ltd	1 730.936	September 1998
John Fairfax Holdings Ltd	1 153.602	30 June 1998
PMP Communications Ltd	1 127.0	June 1998
Ericsson Australia	1 085.976	1997
Tech Pacific	1 062.123	1997
Hewlett-Packard Australia	965.2	October 1997
Independent Newspapers	848.4	June 1998
Seven Network Ltd	803.051	27 June 1998
Fujitsu Australia Ltd	782.5	March 1998
Australian Broadcasting Corporation	706.504	30 June 1998
Digital Equipment Corporation	702.1	June 1998
Village Roadshow Ltd	636.351	30 June 1998
Canon	573.1	December 1997
NEC Australia Pty Ltd	563.4	March 1998
Alcatel Australia	554.6	December 1997
AAPT	508.401	30 June 1998
Siemens Ltd	492.1	September 1997
Hannan Group	490.0	June 1998
Queensland Press	449.4	June 1998
Amalgamated Holdings: Greater Union	394.675	June 1998
Rural Press Ltd	341.338	June 1998
Australian Provincial Newspaper Holdings Ltd (APN)	337.756	30 June 1998
Apple Computer Australia Pty Ltd	230.0	September 1997
Ten Network Holdings Ltd	227.237	31 August 1998
CSC Australia Pty Ltd	195.0	March 1997

Note: Unless otherwise specified, the net annual revenue refers to revenue generated within Australasia during the 1997/98 financial year. The Australian Broadcasting Corporation is included to provide a basis for comparison, though its revenue is predominantly from government appropriation.

Sources: Company Annual Reports; *The Ian Huntley's Shareholder*, 15th edn, Ian Huntley Pty Ltd, 1998; *Australia's Top 500 Companies 1998–9*; Dun & Bradstreet (Australia) Pty Ltd, 1999; *Business Review Weekly*, November 1998; and *Australian Financial Review*, Supplement, May 1999.

Table 7.2 Growth sectors of the Australian economy, 1993/94 to 1997/98

Sector	% growth, 1997/98 (1 year)	% growth, 5 years to 1997/98
Communications	16.6	81.1
Construction	0.3	30.4
Property/Business services	9.3	25.7
Mining	7.8	23.5
Finance	5.9	29.4
Recreation	5.9	24.3
Retail trade	5.3	19.8
Personal services	5.0	23.2

Source: Australian Bureau of Statistics, published in the *Age*, 4 June 1998, Business, p. 1. March quarter national accounts. Trend figures at 1989/90 prices.

is Australia's steel and energy giant, BHP, the big Australian. The data in Table 7.1 show a significant group of major communications corporations trading well in Australia; nine exceed billion dollar revenues and another eleven trade at between $0.5 billion and $1 billion annual revenue.

Data for the Australian communications sector as a whole shows that it grew by an impressive 16.6 per cent for 1997/98 and the sector has almost trebled its output in the past decade. It is now the fastest growing sector of the Australian economy, overtaking agriculture, food processing, metal products, machinery production, electricity, road transport and education. Table 7.2 shows the sector's spectacular economic growth relative to other sectors during the past five years.

Over the past two decades, economic growth in Australia has been rising in terms of GDP per capita, and as Tom Mandeville argues, employment-generating industries 'have been the smart, knowledge-intensive industries of business services, communications and media, tourism, education, health and financial services'. He adds that 'since 1960, business services have not only doubled in importance to the economy, but have increased its employment opportunities four times'. Between 1988/89 and 1996/97 the communications industries in Australia grew at nearly five times the rate of the overall Australian economy, and its workforce grew by 10 per cent during that same period.[22] So, Australia has experienced substantial growth in information-based industries and restructured towards a 'post-industrial economy', although not to the extent that Barry Jones had called for in his 1982 seminal work, *Sleepers Wake! Technology and the Future of Work*. The big question now is still how we might realise more of the benefits of this new communications order in the future.

Convergence brings in a wider range of economic activities, and with it some methodological problems about what ought to be included

Table 7.3 Australian communications industry sector, 1997/98

Industry class	Revenue ($bn)	Five-year growth (%)	Value added (%)	Employment ('000)	Establishment (number)
Hardware, consumables and servicing					
Computer and business machine manufacturing	1.3	−3.5	0.23	2.8	155
Telecom equipment	2.73	1.4	1	8.1	160
Other electronic equipment manufacturing	2.42	5.8	0.88	11.0	400
Computer wholesaling	14.5	4.0	2.2	29.8	2 519
Electronic equipment wholesaling	1	4.5	0.22	3.8	370
Computer software retailing	1	8	0.28	2	600
Computer maintenance	1.6	0	0.4	4.5	398
Peripherals and stationery	1.6	11	0.51	n.a.	n.a.
Software and services					
Recorded media manufacturing	0.21	4	0.14	0.7	15
Computer consultancy	9.81	17	4.44	62.9	11 400
Imported software, other consultancy	2.1	15.5	1.1	1.1	n.a.
Telecommunications					
Telecommunications	22.4	5.2	10.5	75	385
Sub-total (IT&T)	**60.7**	**6.9**	**21.95**	**202**	**17 000**
Multimedia					
Radio	0.85	0.8	0.3	6.9	326
Television (incl. pay TV)	3.47	6	1.87	13.5	428
Video hire	0.7	−1.1	0.17	12.9	3 046
Information					
Data processing services	1.85	16	1.06	7.1	406
Databases	0.25	16.4	0.1	1.3	115
Education and training	1.6	15.5	0.96	n.a.	n.a.
Total	**69.66**	**7.2**	**26.41**	**245**	**23 000**

Note: 'Computer consultancy' includes outsourcing; 'telecommunications' includes the Internet.
Source: IBIS Business Information, June 1999.

in the data. One of the ironies of our information age is that we cannot often get accurate information about the size of the industry, or agree on its scope. If, however, we aggregate the financial data across all of the sectors of electronics, information technology, and telecommunications platforms, carriage and services, in addition to media, information and entertainment services, we clearly have an economic sector of very great significance to the Australian economy. The work done by IBIS Business Information suggests that by the financial year 1997/98, the Australian communications sector had become a $70 billion sector, as detailed in Table 7.3.

The data in Table 7.3 on the growth of communications are indicative of something more profound than merely a new expansionist phase of the Australian economy. In 1997/98 the Australian communications sector had a revenue base of an estimated $57 billion. By comparison, Australia's oldest industry sector, agriculture, forestry and fishing, had aggregated revenue of about $31 billion, constituting about 3.2 per cent of the nation's GDP. This was a major change from its 25 per cent contribution to Australia's GDP at the beginning of this century.[23]

Gross revenue figures inevitably fail to portray a complete picture of the economic value of an information society. How can the various components of information in society really be quantified, and how do we add up the various types of knowledge that exist? As a World Bank report asked,

> What common denominator lets us sum the knowledge that firms use in their production processes; the knowledge that policy making institutions use to formulate, monitor, and evaluate policies; the knowledge that people use in their economic transactions and social interactions? What is the contribution of books and journals, of R&D spending, of the stock of information and communications equipment, of the learning, and know how of scientists, engineers, and students? Compounding the difficulty is the fact that many types of knowledge are accumulated and exchanged almost exclusively within networks, traditional groups, and professional associations. *That makes it virtually impossible to put a value on such knowledge.*[24] [author's emphasis]

A generic economic shift is under way in our society.

Australia as an IT client state

While Australia's overall economic growth in communications in recent years has been impressive, there are some serious structural problems, notably our dependence on a high level of imports in information technology.

Internationally, the IT industry is often portrayed as an industrial economist's dream come true. The history of IT resides in the extraordinary growth of the computer industry with its brilliant entrepreneurs, silicon valleys, and unregulated, free-wheeling product development for ultimate testing in the marketplaces of the world. What is often forgotten is that the development of IT in the world's leading IT nation, the United States, was substantially seeded by the US government in reduced R&D to private companies. Broadcasting and telecommunications emerged out of statism, for good reasons, with governments heavily involved in development and ongoing regulation. The world's IT industry is really arch capitalism in microcosm, symbolised today by a brilliant computer software developer, Bill Gates, rewarded by becoming the world's richest individual. Australia, however, has not done well in its indigenous computer development, remaining highly dependent on the world's big computer brand names as essentially an IT client state. A 1998 stocktake of Australia's information industries argued:

> Strong information industries are strategic to Australia's relative wealth creation capabilities in the 21st century. They are globally large, fast growing and will increasingly underpin activity in most other sectors of the economy. While production and exports by Australia's information industries have grown in recent years, they would need to increase *six-fold* to reach the level such that our share of global production of IT&T products and services matched our share of global consumption.[25]

The IT industry is segmented and includes network equipment, terminal and peripheral equipment, computer equipment, digital equipment electronics manufacturing, semiconductors and software. There are some serious methodological problems in measuring the scope and size of the Australian IT sector. Essentially the term 'information technology' centres on computer hardware and software, although these distinctions are less clear than they used to be, and the lines between information technology and telecommunications are blurred in some functions. In manufacturing, switching systems made for telecommunications carriers are partly electronics equipment and partly computer software, and so are difficult to split for any collective industry revenue calculation. Telstra and IBM Global Services Australia (GSA) have a $4.4 million outsourcing agreement—is this telecommunications or IT? There is also a professional services market including consulting, systems integration and training built around computing which is difficult to quantify. There are over 9000 computer specialist firms employing 55 000 people in Australia, many with close links with telecommunications work.[26]

Several major reports to the Commonwealth government have grappled with these difficulties but highlighted Australia's relative position in IT internationally, and identified some serious national problems. Most notable were the March 1997 report, *Spectator or Serious Player? Competitiveness of Australia's Information Industries*, by consultants David Charles, Roger Allen and Roger Buckeridge, and former Liberal Party President Ashley Goldsworthy's report, *The Global Information Economy: The Way Ahead*, prepared on behalf of the Information Industries Task Force and published in July 1997. Collectively these reports sounded warning bells regarding Australia's economic position in terms of world trade and competitiveness in information goods and services. Despite the extraordinary growth of our information economy in the past decade, our relative trading and competitive position has actually worsened when measured against international benchmarks. There are three areas of great concern; our information technology and telecommunications (IT&T) trade deficits, our poor venture capital markets, and our weak approach to local research and development.

Goldsworthy's report pointed out that Australia has a serious balance of payments problem in IT&T, and is an insignificant player in global information and communications technology trade, responsible for a mere 0.3 per cent of world exports in office machinery and telecommunications (1995). While Australia's exports of information and communications products and services have increased in recent years, our relative IT&T import dependence has become far greater. For the period 1990–95, information and technology equipment exports increased from $772 million to $2.2 billion, while related service exports increased marginally from $739 million to $880 million, but our comparable trade deficit increased dramatically from $3.4 billion in 1990 to $6.7 billion in 1995.[27] Moreover, Australia paid over $17 billion in 1995 to overseas suppliers of product and intellectual property, which included video and television programs.[28]

Buckeridge and Allen also argued that the trade in IT products and services was running four to one against Australia, and they estimated the deficit 'was $14 billion in 1995 and is likely to go on widening and could exceed $30 billion by 2005'. Alarmingly, the Goldsworthy/Information Industries Task Force estimate was even more pessimistic, arguing that 'projections of Australia's information and communications technology trade trends during the 1990s suggest that the deficit could reach as much as $46 billion by 2005!'[29]

Underpinning the balance of payments problem is Australia's weak venture capital market. Innovation in high technology needs dynamic capital markets, a factor of great significance to those economies with booming IT industries. However, the World Competitiveness Report, which publishes an annual scoreboard of national performance, placed Australia fourteenth out of 48 countries surveyed.[30] Australia ranked twentieth of 46 countries for access to venture capital and 37th out of 46 countries for entrepreneurialism and innovation.[31] Also, David Mortimer's more generic industry report, *Going for Growth*, published in June 1997, pointed out that Australia's share of regional foreign direct investment has fallen from 37 per cent in 1980 to less than half that figure in 1997.[32] Banks and superannuation funds in Australia's local capital markets have tended to avoid strategic investments in the IT field, preferring safer stocks and shares and traditionally more secure investments in real estate. Australia urgently needs to reconfigure its economy towards value-added industries, yet it appears to be virtually impossible to kick-start major strategic investment projects through the existing Australian venture capital markets.

Also, in terms of research and development, Australia has long been a technological client state in information technology. The computer industry is dominated by the world's great foreign brand names, who essentially regard Australia and the rest of the world as little more than a marketplace, and who fund very little local R&D wherever they operate and market. Australia ranked seventeenth among 22 countries in terms of the level of business expenditure on R&D as a percentage of GDP.[33]

So much for a depressing trilogy of indicators of poor national performance of Australia's IT: bad balance of payments, weak venture capital markets and an inadequate local R&D performance. Our performance in terms of indigenous development has been much more impressive in media, broadcasting and telecommunications than it has in IT.

A profound change is under way in the nature of Australia's industrial complex, part of which appears to be the decline of the established industrial–manufacturing base in which the manufacture of telecommunications equipment in Australia was important. We are clearly better at being consumers, rather than suppliers, of communications goods and services. What does this mean for Australia's future? If manufacturing is in a state of decline, what is the appropriate policy response? Can we do anything about its decline, or doesn't it matter anymore? Is it credible to argue that this traditional industrial base will merely be replaced by a service-based society of comparable productivity?

Michael Porter's seminal work, *The Competitive Advantage of Nations*, argued that a globally competitive manufacturing sector is a primary requirement for any modern nation. High-technology goods and services are the fastest growing areas of world exports. Between 1985 and 1992, world exports of high-technology products grew by 15.6 per cent per annum, faster than services exports at 11.1 per cent and faster than merchandising exports as a whole at 11 per cent.[34] In 1993/94, manufactures as a whole earned more income than any other sector of the Australian economy—more income for Australia than primary produce or mining.[35] For the first five years of the 1990s, Australia's electrical and electronic industries were the two fastest growing economic sectors, with annual export growth of 32 per cent for telecommunications equipment compared to an overall average of manufactures of 12.6 per cent. In 1995 the absolute productivity level of manufacturing was 15 per cent higher than for Australian industry as a whole, and the communications sector supported wages 30 per cent above average weekly earnings. Clearly, knowledge-based goods and services are the fastest growing in terms of output, employment and wages, trends that are likely to strengthen in future in view of the growth of the information economy and world trade liberalisation.

Yet in terms of the production of *goods*, there are some alarming signs in the late 1990s of structural change in Australian manufacturing in communications. Alex Gosman, Executive Director of the Australian Electrical and Electronic Manufacturers Association (AEEMA), summarised the bleak manufacturing picture, pointing out that 'Australia has virtually no electronic components industry and no large scale computer chip manufacturing. There is virtually no materials supply sector, printed circuit board makers are small and losing business, electronic assemblers are under pressure and transnational investment is being lost.'[36]

In 1998 the two transnationals with the biggest telecommunications equipment operations in Australia during the past 30 years, Ericsson and Alcatel, abandoned manufacturing in Australia. Similarly, IBM's Wangaratta plant, originally a manufacturer of typewriters but which became IBM's South-East Asian computer manufacturing hub in the 1980s (and thereby the most significant computer manufacturing operation in Australia), was sold to newcomer contract manufacturer Blue-gum Technologies. The Wangaratta plant contributed $611 million to the exports of IBM Australia in its final year, but Managing Director Bob Savage said that manufacturing was no longer one of the company's 'core competencies' and that proximity to markets was a major reason behind the decision to sell. However, Peter Roberts of the *Australian Financial Review* suggests that the trend among domestic companies that match intellectual property development with communications manufacturing is dismal: JNA Telecommunications, Datacraft and Eracom were recently sold overseas, joining AWA, Stanilite, Exicom, Dataplex and Netcomm which have either been bought out or closed.[37]

Why has this happened? The short explanation resides in generic changes in the nature of telecommunications equipment in an increasingly intensely competitive globalised electronics manufacturing industry. Briefly, whereas in the past the telecommunications common carriers required generational changes to hardware, such as the major changeover from analogue-based networks to digital networks in the

1970s and 1980s which provided great continuing business for electronics manufacturing companies, many of the present modern telecommunications networks no longer generally require substantial comparable hardware upgrades. Since the early 1990s the mix between hardware and software network components has changed. Instead of Telecom Australia regularly ordering completely new electronic telephone exchanges from its preferred manufacturers, Telstra Australia has new high-capacity networks (with far fewer switches) that are more permanently in situ, but which can be made more flexible and essentially upgraded to offer more services through changes in software.

The Telecom networks were 90 per cent telephony networks until the 1990s, although Telecom provided major data networks from the early 1980s—Austpac, DDN, later ISDN and Fastpac. In the mid-1990s Telstra had a $3–5 billion hybrid fibre coaxial cable program to provide for pay TV access which it curtailed after about $2 billion had been spent. More recently, Telstra has had to shift focus towards the rapidly overtaking Internet voice and data traffic, part of *reorienting itself from its origins as a telephone company to become an information service corporation.* Gerry Moriarty, Telstra's group Managing Director for network technology and multimedia, has pointed out that in 1994 the Internet traffic represented just 1 per cent of international voice traffic, but in 1998 it exceeded voice traffic on many routes, having grown by a staggering 10 per cent a month for the past four years![38] This explosion of traffic has meant that Telstra is now looking at the biggest upgrade of its network since its $3.3 billion program, Future Mode of Operation, 1993–99, to convert the core telephone network from analogue to digital technology. Essentially the new upgrade will involve the installation of software and other infrastructure (for example, high-capacity routes) replacing telephone exchanges to support Internet protocol services such as voice traffic and multimedia over the Internet. It is the software engineers rather than the electronics manufacturers who have come to occupy the centre stage of future telecommunications infrastructure.

Similarly, until the late 1990s, the large business customers tended to purchase a private automatic branch exchange (PABX), an off-the-shelf 'phone box' which interconnected with the carrier's public switched network, and which generally offered uniform telecommunications services within the company from the one site. Since the late 1980s, PABXs were designed for the transmission of both voice and data, intercom and paging functions, speed dialling, call forwarding, message waiting, voicemail services and conference calls, all across multiple sites. What has made such a revolution in service functions possible is the extraordinary development of intelligent network software which represents a focal shift away from traditional hardware-based telephone networks to the new multifunctional intelligent software networks of the future. *The shift to software opportunities must therefore be central to strategic thinking about Australia's development prospects in communications.*

These major changes are impacting on traditional telecommunications manufacturing businesses, and established corporations have adapted their strategic thinking towards the software/services revolution. The world's US$250 billion equipment industry is now consolidating around mergers between traditional network manufacturers and those players who can offer cutting-edge services in new flexible, intelligent networks. In 1998, for instance, Nortel, North America's second-largest equipment provider, acquired Bay Networks, a leader in Internet protocol technology; Alcatel of France, another established traditional equipment manufacturing corporation, bought DSC Communications, a core intelligent network equipment provider; and traditionalist Tellabs bought Cienna, a leader in optical network technology. It appears that only a small number of equipment manufacturers will have the capital clout, technology scale and global reach to survive in this new telecommunications jungle. A 1998 *McKinsey Quarterly* summarised a growing perception that 'US observers are already talking of a "big three" in data networking: Lucent, Nortel, and Cisco Systems. To that list add European competitors such as Alcatel, Siemens, Ericsson, and Nokia. These three titans—each of which have IP-based packet switching network technology at the heart of their operations—will have to fight multiple battles.'[39]

And what does this mean for Australia? There is growing concern in industry and policy circles that Australia may not have a significant telecommunications manufacturing future, but there is no consensus on how we might replace this vital source of productivity and mode of employment. It is important to recall the serious level of predicted trade deficits before the more recent departure from manufacturing of the companies cited above. Buckeridge and Allen pointed out that the trade in IT products and services was running four to one against Australia and they estimated the absolute deficit 'was $14 billion in 1995 and is likely to go on widening and could exceed $30 billion by 2005'.[40] The Goldsworthy report pointed out that 102 semiconductor plants are currently being built or planned in the world, on average involving 1000 to 2000 jobs at more than $700 million in output, with 62 of them in Asia. None of them is planned for construction in Australia.[41] Alarmingly, the Goldsworthy/Information Industries Task Force reports' collectively estimated deficit was even worse, based on ABS projections, stating that 'Australia's information and communications technology trade trends during the 1990s suggest that the deficit could reach as much as $46 billion by 2005'![42]

In the aftermath of these reports and projections there was considerable debate about what we might do about such an obviously serious national situation. There was a short period when forms of policy intervention were proposed in several quarters. In the period just before the Asian economic downturn in 1997, the Singapore-style corporate tax exemption for 'pioneer firms' for ten years, and Malaysia's investment tax allowance of a write-off of 100 per cent of capital expenditure in the first five years, were often proposed as measures to be taken seriously by the Australian government. Also, following the release of the Mortimer report the government appointed former Optus and Fairfax boss, Bob Mansfield, as the government's strategic investment coordinator, and the prospect of a $1 billion government fund was floated to provide incentives to attract key manufacturing investments to Australia.

A policy test case emerged concerning the prospect of the government offering financial investments to lure high-profile silicon chip manufacturing plants to Australia. The Israeli government had paid $600 million to attract Intel, the largest chip fabrication manufacturer in the world, to establish an Intel plant in Israel. In April 1998, Intel President, Craig Barrett, who travelled the world negotiating with governments, presented what he described as a 'digital deal' to Australian Prime Minister, John Howard: commit to a financial package that would bring an Intel $6 billion computer chip factory to Australia, or walk away with nothing. According to Barrett, '[T]here are countries around the world that will give you a 40% capital rebate on your investment, so if you're going to make a $US2 billion investment, that's a $800 million grant back to you.'[43] No deal was ever sealed with the Australian government, but as the international financial crisis deepened, serious doubts emerged as to whether Intel itself would then commit to any new plant anywhere.

In 1998, industry and policy consensus emerged around these debates about 'industry policy' that Australia had to put in place certain industrial fundamentals that would enable it to compete on its own terms in an intensely competitive global marketplace. Create an environment where new projects were viable for all, not just for specific industries or particular investors, the argument ran, rather than providing financial investments in the form of handouts from government, and put the right incentives in place for investment. However, laborious debates about generic tax reform have more recently subsumed issues about the best industry policy for Australia. Local manufacturing is in for a tough time. The Economist Intelligence Unit estimated in 1997 that over the three-and-a-half years to the end of year 2000, between 85 000 and 100 000 full-time jobs across the one million full-time workers employed in Australia's manufacturing sector are at risk.[44] For computer and telecommunications equipment, Senator Richard Alston, Minister for Communications and the National Information Economy, was reported as saying that he considered 'Australia had missed the boat on hardware'.[45] Not even former ACTU President, Simon Crean, Labor's industry policy spokesperson, proposed an interventionist 'save local manufacturing' policy during the 1998 election campaign. Australia now has a new industry policy—it is called global competition.

Perhaps we must turn to new sources of vision to build a different kind of Australian industrial complex for the future.

8 Whose vision?: Third way communications

> *Our competitors in the global economy are already taking significant steps to embrace the online economy and, as first movers, may capture an important advantage . . . Winning online nations will enjoy lower costs and higher service exports, leading to employment and strong national economies.*
>
> australia.com: Australia's Future Online, Australia's Coalition of Service Industries, Melbourne, 1997, p. 7

> *It promises the world, but what has the Internet really delivered so far? There is E-mail and porn, but scant sign of a commercial or cultural revolution. Business is scratching its head, educators are questioning the benefits, and many people are wondering whether they have been conned.*
>
> Helen O'Neill, 'The Internet backlash', Weekend Australian, 15–16 March 1997, Syte 5

American President Bill Clinton and British Labour Prime Minister Tony Blair have popularised the notion of the 'third way' as a useful, though contentious, framework to examine contemporary society. 'Third way' thinking has been articulated by Robert Reich, former Secretary of the US Department of Labor, and Anthony Giddens, Director of the London School of Economics and Political Science. In linking the shifts in international political economy to communications, a fundamental theme of this book, we now need to examine the relationship between changes in communications and the notion of the 'third way'.

Briefly, the period of the 'first way', from the post-Second World War era until the early 1980s, was when the Keynesian welfare state was the dominant Western ideology. The centrepiece of the 'first way' was a balanced partnership between the state and the market. The private sector was well rewarded, but its role in the partnership was to achieve economic growth for the nation in order to maintain full employment. Henry Ford, it was said, raised his workers' wages so that they could buy more cars. The role of the public sector was essentially to establish publicly owned utilities for those services not provided by the private sector, and to distribute the rewards of economic growth more equitably through 'cradle to grave' welfare state programs. There were generally agreed boundaries between the public and private sectors, although it was widely accepted that governments should intervene to correct market failures and malpractices through regulation.

The 1980s saw the emergence of the 'second way' when the Keynesian model was essentially replaced by the model of 'economic rationalism' in most Western democracies. This holds that governments should fundamentally reform the national economy, as well as their own practices, with market concepts of efficiency, privatisation, deregulation and wholesale competition. Governments should be 'slim', intervene less, promote 'flexible' labour relations, and change the former comprehensive system of state welfare to provide safety nets only. This is a society based on market fundamentalism and economic individualism, with wider acceptance of inequality throughout society and generally with much less compassion for fellow human beings.

In the 1990s the 'third way' has emerged from a widespread belief that economic growth is spurred by free market policies and free trade in a global economy. Reich argues, 'Don't try to block change; don't protect or subsidise old jobs in old industries, and don't keep people who aren't in work on the dole. Embrace economic change, but do it in a way that enables everyone to change along with the economy.'[1] For Anthony Giddens there was, first, the classical social democracy (the old left), and second, Thatcherism or neoliberalism (the new right), but the third way is seeking 'the renewal of social democracy' which gains benefits from the new era of global market forces. For Giddens, the third way, much less clearly defined,

> . . . refers to a framework of thinking and policy making that seeks to adapt social democracy to a world which has changed fundamentally over the past two or three decades. It is a third way in the sense that it is an attempt to transcend both old style democracy and neo liberalism.[2]

These are not three clearly delineated political orders, and the notion of the third way has yet to be unpacked in concise terms. Giddens points out that the cynics see Blair's new Labour as persisting with the economic policies of Margaret Thatcher, and that Clinton talked of social justice yet also said there would be an 'end to welfare as we have known it'. In Australia some members of the Labor Party are third way converts, notably Mark Latham, but also Lindsay Tanner, who has argued:

> All over the world, people on the left of politics are grappling with the profound implications of fifteen years of cataclysmic change . . . I hope this book will assist those who believe in community and society to adjust the means by which they pursue the unchanging objectives of social equity and economic justice in a radically different world. We do not have to choose between Old Left and New Right. There is another path.[3]

How might this new path be articulated in the context of the major recent changes experienced in Australian communications? To bring together some of the key arguments presented in this book, and to move forward to possible action plans for the future, a third way of communication policies and practices is offered.

Australia's communications environment is in a state of major transition and there are few signs that the scale and rapidity of change is likely to abate. Australia's new world of communications is evolving with a complex set of political, economic, social and technological forces driving changes that are not easily interpreted, but we must attempt to understand these changes in order to benefit from the major structural changes occurring in contemporary society. *In short, Australia needs to ensure that it devises strategies to develop an advanced networked society.* The term 'network' means not only the physical infrastructure for the transmission of information but *connectedness for purposes of communication between people* in ways not possible before.

There is the real prospect that we can enrich the way that many people communicate with each other. Yet how do we progress towards this desirable goal? How does Australia build an advanced networked society? Whose responsibility is this, what do we need to do and how do we do it?

The third communications order: e-communities/ e-Commerce

The first communications order (up to the late 1980s) built a national network to basically provide telephone services to almost all Australians. The second order (from the late 1980s to the late 1990s) was driven by a sense that we would 'fall behind' other major democracies and our trading partners if we did not embrace competition, privatisation and deregulation as the driving forces of national economic policy, including telecommunications. The new communications order (post mid-1990s) extended the competition paradigm with a range of complex variables, such as multiple stakeholders, multiple interconnected networks, intensified competition, and an explosion of possible information and communications services. But this remarkable energy in the marketplace has also been linked to poorly defined markets, users polarised by their capacity to pay and by geographic restrictions in access to advanced services. We need to see the present changes as part of a new communications order with opportunities for doing business with consumers in new ways through electronic commerce, *together with* the new modes of communication that the Internet brings to citizens, as e-communities. *The essence of the third communications order is to expand our global opportunities through electronic commerce with new networks, and also provide Australians with the benefits that can come with membership of electronic communities.* We must become better at thinking and acting both locally and globally. In a nutshell—e-communities/e-Commerce. We need to develop our strategic thinking a great deal more than we have to date before any credible third way communications order can emerge.

What ought to be driving forces in the development of a third communications order? Nowadays it is common to hear calls pleading for visionaries to take us into the future, usually expressed as our 'journey into the next millennium'. In the context of the national development of Australia's information economy and society, the fundamental question that arises is who are the visionaries, where are they coming from, what is their sense of vision, and what is the best strategic course of action for Australia? Is the vision intended to serve the economy, business interests, consumers, citizens or governments, or could it serve the nation as a whole? Or is it pointless to talk about vision, which implies setting priorities and targets for desirable outcomes?

There is a plethora of visionary positions taken on the future of Australia in the context of an information society. In an important symbolic gesture in September 1997, the Prime Minister, John Howard, announced the creation of a new National Office for the Information Economy (NOIE) within the portfolio of the Minister of Communications and the Arts, Senator Richard Alston. NOIE was later to issue its mission statement in preparing Australians for the information economy: *'To ensure that the lives and work of Australians are enriched, jobs are created, and the national wealth is enhanced, through the participation of all Australians in the growing information economy.'*[4] How might such commendable objectives be achieved? In a world of shifting power bases, what are the respective roles of the public and private sectors in the emergence of a new economic development paradigm? How can we create businesses in Australia—large, medium and small—which can trade successfully in the new global economy? Where is our thinking heading about opportunities for access by all Australians to our new era of communications?

Well before the existence of NOIE, some state governments had produced their own sense of vision. The Victoria 21 strategy is widely regarded as one of the most visionary of any Australian state government to date. In the document *Victoria: A Global Centre for the Information Age*,[5] Premier Jeff Kennett suggested that 'Victorian firms will carve out a major role in the global provision of multimedia products and services, and the government will use the power of communications technology and multimedia to transform the way it provides services to and communicates with the public.' This kind of thinking articulates an approach to development responsibilities which might form part of the basis of a third way: *government with innovative policy leadership and as a major user of information technology, together with private sector investment and rewards, especially where successful in global markets.*

Where does the private sector stand on these issues, and what is their sense of vision for an advanced information society? One statement of vision for the development of communications in Australia constructed with private sector interests and responsibilities firmly in mind that we might embrace came from the private sector global management consultants, McKinsey & Company, in a report to the Australian Coalition of Service Industries, titled *australia.com*. It argued that 'businesses create individual, industry-sector and national visions for the online economy in important areas such as health care, financial systems, logistics, etc. to ensure activity and competition are focussed on building value for the nation. Governments actively reinforce this national vision—aligning state, federal, and local government objectives to avoid dysfunctional discrepancies across jurisdictions—by actively promoting the use of online services throughout the community.'[6] According to McKinsey, *the development process ought to integrate both private sector and public sector vision and practice.* Their argument is that the private sector alone cannot undertake such a radical restructuring of business in society and promote the benefits of on-line services to the wider community.

Terry Cutler, Australia's most notable private sector on-line luminary, has pulled together a constructive set of objectives for consideration across many different threads of communications development. He offers a ten-point agenda for Australia to be successful within the global information economy:

1. Build our national skill base.
2. Get our telecommunications fundamentals right.

3. Grow innovative on-line content and programming.
4. Internationalise all our activity.
5. Exploit the new information economies.
6. Create Australia as a safe, efficient port for trade in intellectual property (that is, get the virtual waterfront right).
7. Import, not export, consumption.
8. Implement a true convergence policy.
9. Enable all Australians and all of Australia to participate.
10. Keep our focus on the economic wood and the user environment of electronic interfaces, not the technology trees.[7]

A valuable complementary perspective to this one emerged in 1996 from a policy forum moderated by the Centre for International Research on Communication and Information Technologies on the development of an Australian on-line society, involving people with different backgrounds and interests, vested and otherwise. They concluded that:

> Australia should aim to create an adaptive climate for the innovation and uptake of information and communication services (ICS). All sectors of society can benefit from Australia building on its present strengths in ICS. Australia's ICS industries must have the capability to meet and encourage demand for advanced communications services thereby encouraging economic growth, job creation and increased productivity. The focus ought to be on sustainable high profit sectors by concentrating on our strengths in market differentiation and recognising ICS as an enabler in key information intensive activities. By enlarging our share of the global market we can increase wealth and societal well being for Australia.[8]

Users of the new networks

Third way communications ought to focus on all of the people who will be involved in this new era of communications. The most fundamental change of all must be *the shift from supply-led development to a demand-led consumer user and citizen participation focus*. In the first telecommunications order it was essentially possible to roll out a telephony network and find that access to a telephone service had a natural take-up as it progressively became an invaluable social instrument. However, the present communications development paradigm has a plethora of information and communications services, of different kinds, requiring difficult judgments about their possible acceptability by users and consumers. There are simply no guarantees for investors and suppliers that the emerging on-line services—such as home shopping, home banking, video on demand—indeed, the generic category of electronic commerce services, have substantial market pull. There is likely to be a great deal of experimentation with new information and communication services, with plenty of market failures and some successes; eventually a sense of order may emerge around these new services. At the end of the day, however, the only services that will work on-line are those that are wanted and are affordable.

The development of so many innovative communications services will require organisations who invest in new communications services to undertake greater investigation of people's needs and greater understanding of the way they run their lives. There are few signs at present of this kind of fundamental research taking place anywhere in Australia. This point needs elucidation, because the eventual outcome and value of the third communications order rests on this central proposition.

Participants who wish to offer services need to *understand user perspectives* rather than technology perspectives. As US Federal Communications Commissioner, Andrew Barrett, has explained, 'When I ask telephone people, "tell me what new services you are talking about", they go into these broad generic terms like telecommuting, telemedicine and distant learning, and I have to remind them that these are not *service offerings*, they are potential *uses* of a broadband network.'[9] Providers of technology tend to start from a position of what they do and what they can offer, instead of starting from trying to understand how consumers might want to use technology-based services, old and new, in their lives. Brenda Dervin explains, 'Almost all our current research applies an observer perspective. We ask users questions which start from our worlds, not theirs: What of the things we can do would you like us to do? What of the things we now offer you do you use? . . . The difficulty is that the data tell us nothing about humans and what is real to them . . .'[10]

The extension of this position is that a great deal more work has to be done in Australia if we are to invest intelligently in services that are going to be useful and needed by Australians, and much more systematic investigation is demanded into social behaviour than hitherto. Supriya Singh, Senior Research Fellow at CIRCIT, explains that 'instead of information being seen as a commodity that has to be transmitted, information has to be seen as a process of sense making where information is constructed by the user. This changes the questions that are asked. Instead of focussing on the service, the questions focus on the customers' needs.'[11] Regrettably, many of the service providers, actual and potential, in this new information services marketplace tend to see this line of argument as 'academic', or lacking commercial reality. On the contrary, however, the successful participants in this new order will be those who understand the social and cultural dimensions of the users in their new businesses.

Assessment of user needs and services will be vital to the success of the third communications order, but there must also be an *appropriate network infrastructure in place to carry and support new services.* The basic transport for telecommunications services may be provided by one or more of the existing public networks, such as the Public Switched Telephony Network (PSTN), the Digital Data Network (DDN), the AUSTPAC packet switching network or the Integrated Services Digital Network (ISDN). However, as the demand for telecommunications services continues to explode, there is constant pressure on the capacity and flexibility of these existing networks to accommodate the burgeoning range of new services. *A vital component of the construction of an Australian advanced networked society is to make affordable high-bandwidth telecommunications available, preferably to all homes, institutions and offices.*

Telecommunications networks have had to respond to a new value chain in services in recent years and this shift is summarised in Figure 8.1.

This is a simplified version of the network of the third communications order. Here the value chain (the four boxes at the top) has radically altered in an environment which has allowed for open competition and unrestricted entry for new carriers from 1 July 1997. The traditional *institutional* distinction between content and carriage has blurred with convergence, and carriers are now involved with content provision (for example, Foxtel subscription television, a Telstra–News/PBL alliance), packaging and publishing services which involve carriers in greater

Figure 8.1 The changing environment

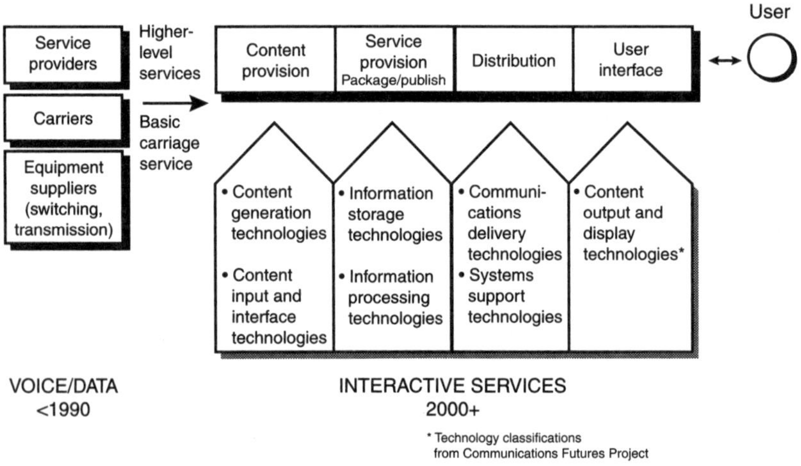

Source: CIRCIT.

Figure 8.2 Possible future broadband services

Government
- Electronic form lodgment
- Public access to government information
- Electronic voting

Education
- Remote interactive teaching
- Virtual classrooms
- Access to overseas libraries

Business
- Electronic commerce
- Multimedia communications
- Videoconferencing

Health
- Video consultation
- Medical records transfer
- Remote monitoring of outpatients

Services to the home
- Video on demand
- On-line banking, shopping, gambling
- Interactive multimedia (eg video games)
- Telecommuting
- Video mail, video phones
- Security services

Community information
- Bulletin boards
- Electronic libraries and museums
- Service directories

Source: Adapted from *Networking Australia's Future*, BSEG, 1994.

marketing functions than before, and dealing with alternative user interfaces (personal or networked computers, television). Their traditional responsibilities of distribution (third box in the value chain), managing the platform delivery technologies, remain as core business. This is no longer the value chain of just a telephone business; rather, it represents a major shift towards a more diversified information services business.

Figure 8.2 shows many examples of possible future broadband services across many facets of Australian society—government, education, business, health, services to the home, and community information.

Universal service opportunities

There needs to be reconsideration of the issues of equity and affordability in this new on-line world. Whereas established media (press, radio, television and telephony) have become largely accessible and affordable, the new media forms of the on-line world (home banking, home shopping and electronic commerce) require more expensive equipment purchases, such as a personal computer and a modem, telecommunications access fees for the Internet, together with the usage-based charges of the Internet service provider. Apart from these matters of user affordability, there are other barriers to user participation. There is the issue of computer literacy which can lock out many potential users who lack the technological skills to become part of this new on-line world. The position of rural and remote people in the context of equity has also received considerable attention, especially since so many rural people have lost local access to many other basic services, notably education, banking, postal and health services, in recent years. Whereas the first telecommunications order was underpinned by a social equity policy of universal services at affordable costs, the new communications order is part of a user-pay society. Unless this issue is addressed in communications policy terms, there will be an increasingly polarised Australian information society with seriously disadvantaged groups, including those in rural and remote communities, the unemployed, those with low incomes and people with disabilities.

The challenge of the third order is not only to accommodate the newly emerging markets, and to serve business and urban residential customers well, but also the full range of other communities, including remote and rural Australians. We need to build digital interactive networks, readily accessible and affordable, with national coverage for the on-line society. However, there is no national policy clarity about the means of achieving this.

Telecommunications infrastructure for the future has been the subject of considerable public policy investigation in recent years. Some important thinking emerged in a report commissioned by the Minister of Communications and the Arts, *Review of the Standard Telephone Service,* published in December 1996. This report advocated that a 'digital data capability' should be 'reasonably accessible to all Australians on an equitable basis wherever they reside or carry on business by 1 January 2000'. This review suggested that the digital data capability should provide a platform for access to a range of services such as fax, e-mail, access to the Internet, electronic commerce and educational applications. There would be significant benefits, they added, for consumers in rural and remote areas in terms of reduced connection times for telephone services and improved monitoring of quality service issues, such as congestion. A carriage service providing such a digital capability should be prescribed, they recommended, under proposed changes to the *Telecommunications Act* from 1 July 1998 'to ensure that it is reasonably accessible to all Australians . . . by 1 January 2000'.[12]

The desirable infrastructure change here essentially involves a shift from networks that provide national access to telephony, to upgraded networks that will provide national access to data, for Internet and other on-line service access. Technically the review panel argued that we need to upgrade and extend the national telephone system to a 9600 bps capability for modem-based data services. Surprisingly, there was little common agreement about the capacity of our existing telephony network in this context; an estimate by Roger Buckeridge was that 32 per cent of Telstra's lines, or 32 per cent of its 2.5 million country, rural and remote services, cannot be guaranteed a 9600 bps bits per second or faster modem data capability, a figure which Telstra contested as an overestimate not based on empirical data.[13]

However, the group could not agree on justification for the level of investment in desirable infrastructure upgrades. Within the ten-member majority of the Standard Telephone Report, a minority report was submitted by academic economist Professor Henry Ergas. Ergas said that the report was 'a disappointing document' where the recommendations, including the central recommendation in regard to a better digital data capability were 'poorly thought through'. He suggested that the capital cost to implement this was unjustified because it was 'likely to run somewhere between $1.6 and $2.5 billion', though, he added, this underestimated the economic cost because it failed to take into account the impact this might have on charges for the standard residential telephone service.[14] Interestingly, the other nine members who supported the proposal included representatives from all three carriers—namely Telstra, Optus and Vodafone—and the Managing Director of the Australian Telecommunications Users Group. However, the minority report scuttled any prospect of action and Henry Ergas carried the day. The government did not act upon the overwhelming group majority recommendation to give Australians a mandated digital data capability by 1 January 2000.

The issue, however, did not go away. In the light of pressure surrounding the possible full privatisation of Telstra, especially in the lead-up to the 1998 federal election, demand for guarantees of 'looking after the bush' simply had to be addressed, especially by a government with rural electorates as part of its natural constituency. Senator Alston, Minister for Communications, Information Technology and the Arts, directed the Australian Communications Authority to hold a public inquiry into 'whether a carriage service that provides digital data capability broadly equivalent to 64 kilobits per second (kbits/sec), comparable to the capability provided by a basic rate Integrated Digital Services Network (ISDN) service, should be incorporated into the Universal Service Obligation (USO)'.[15] The ACA report, tabled within four months, argued that there were 'encouraging signs' that the data services market would deliver the capability to address the increasing demand evidenced by Internet and e-mail statistics. They added, however, in a masterly understatement, that despite promising signs, 'disparity exists in terms of data service capability and access charges between metropolitan and rural services'.[16]

The ACA's analysis of the current telecommunications infrastructure in Australia, and its capacity to provide data services, showed that the benefits of competition policy were most evident in metropolitan areas. In summary:

- The Telstra PSTN is currently capable of providing data services at a data rate of 9.6 kbits/s to 95 per cent of its urban and major provincial subscribers, but to only 70 per cent of its rural and remote subscribers.
- The current capability provided data services at a data rate of 28.8 kbits/s to 60 per cent of urban and major provincial subscribers, but to only 30 per cent of its rural and remote subscribers.
- Metropolitan subscribers are afforded greater choice of carriers and Internet service providers with access to ISPs at an untimed local call rate.
- Conversely, in many rural areas, Telstra remains the sole provider of data services through their standard telephone service, and subscribers are restricted in their choice of ISP.
- Of the total population of rural and remote subscribers, about 37 000 do not have access to untimed local calls.

One of the variables that the ACA had to come to terms with in framing their recommendations was the plethora of terrestrial and space systems under development which will enable narrow band and wider broadband infrastructures to provide new telecommunications services. In particular, forthcoming satellite systems (LEO, MEO, VSAT) will offer a wide range of new services, including simple messaging, e-mail services and non-voice small Low Earth Orbit systems ('little LEO') to narrow band voice and higher speed data (big LEO and MEO).[17] Undoubtedly, carriers will use a hybrid of technologies in the future to meet demand in both urban and rural areas for the delivery of data services, and they will need to be able to exercise a great deal of discretion in developing new communications platforms for the future.

The principal recommendations of the 1998 ACA Digital Data Inquiry were:

- The costs to the community of specifying, as part of the USO, a carriage service broadly comparable to a digital data channel with a data rate of 64 kbits/s to end users as part of the designated basic rate ISDN service *would outweigh the benefits to the community* if it were provided solely by ISDN.

- On the basis that Telstra's proposed satellite access service meets the criteria specified in section 141(2) of the *Telecommunications Act 1997*, then the Australian community would likely not incur additional costs as a consequence of a service broadly comparable to a digital data channel with a transmission rate of 64 kbits/s being specified as part of the USO.
- An extension of Telstra's current licence condition to provide ISDN/64 kbits/s to 100 per cent of the Australian population would have a significant commercial impact upon Telstra.

There was perhaps a predicability of outcome for this inquiry given that they were given such narrow terms of reference: essentially its members had to respond to questions rather than recommend the best policy. Moreover, the overriding paradigm was competition policy, and there are now many more delivery options open in an environment with several major carriers who inevitably must take responsibility for their own investment decisions. The days of mandated telecommunications social policy fit uncomfortably with this new highly deregulated, intensely competitive marketplace. The ACA pointed out that no country has specified a data rate capability as part of USO arrangements, apart from the inclusion of ISDN in USO arrangements in Germany, Denmark and Norway.

The whole set of arguments about USO costs and infrastructure responsibilities for the future was subsequently thrown into confusion. In the search for some solutions to the problems of assessing the cost of USOs, all industry players had agreed with the ACA that a new model should be adopted. The US industry consultant Bellcore subsequently devised a new model which surprisingly revealed that roughly 494 000 rural customers were unprofitable and being subsidised, as opposed to the former BTCE estimate of 55 000. Telstra, which meets about 85 per cent of these costs, was quick to point out that this meant that the total cost of the USO was nowhere near the estimated figure of $251.6 million, but of the order of $1.8 billion for 1997/98! Telstra argued that while it is now fully exposed to market competition, these costs are 'largely unfunded by government and other industry participants'.[18] However, other carriers would face substantial cost increases if they were ever required to make their contributions to the USO fund based on this revised figure. For Cable & Wireless Optus, for instance, their $24 million contribution to the cost of the USO in 1996/97 would rise to an estimated $180 million if the figure of $1.8 billion was accepted. On 12 October 1998 Senator Richard Alston announced that the government would seek to reach agreement with all telecommunications carriers to cap the cost of the USO at $253.32 million for the 1997/98 year, and if such an agreement could not be reached, then the government would legislate to cap the USO cost for 1997/98 at that level. Subsequently, the Minister directed the ACA to report by 31 March 1999 on the validity of Telstra's $1.8 billion Net Universal Service Cost (NUSC) claim in providing telephone services across Australia and this was set at $580.2 million in July 1999.

No cost-benefit analysis could ever prove that predicted growth in new telecommunications services would justify the substantial levels of infrastructure investment required for an ambitious level of national data capability to all homes, offices and institutions. The capture of the policy agenda by today's economists means that we will no longer make long-term public policy strategic decisions about future infrastructure as we did with the first communications order. We have now set in place a competitive telecommunications policy regime that will produce a range of carrier organisations, differently owned and organised, vigorously competing with each other, with varied interests in network development. Although competition policy is likely to produce a more dynamic marketplace, with many players offering many services, it will not necessarily create the third order national infrastructure network that is needed, one that is accessible to almost all people in Australia. Clearly, there was no Community Internet Plan in the 1990s comparable to the 1960s Community Telephone Plan. In fact, there are many disincentives for carriers, especially Telstra, to build the third order network that is needed. Competition policy will inevitably mean that their locus of commercial energy will be directed towards what they perceive as the most profitable components of their business, as their new shareholders will demand. Infrastructure development must necessarily involve a range of stakeholders, each with different interests for different regions with different services, but the overall national policy must still maintain the principle of equity—universal services at affordable costs for the information age.

A compromise school of thought on infrastructure development has advocated the notion of 'universal reach'. The Broadband Services Expert Group argued in 1994 that 'it is not financially feasible to provide interactive broadband services to all Australian households at this time'. Interactive services which provided, for instance, video on demand to the home, would be a substantial further upgrade on what is proposed above. As an alternative, argued BSEG, 'the group proposes a concept of *universal reach* as a way of describing how to make broadband services available to as much of the community as possible, as quickly as possible. This approach underpins our strategy to provide access to the community in the near term.'[19] The notion of universal reach essentially advocates a higher level of institutional access for new information services rather than planning for ubiquitous household access. This could also involve fast tracking new networks to support specified community-based initiatives, as has happened in Canada with Nortel and New Brunswick Telecom. Hence, for universal reach there would be better access to many information services in places such as educational institutions, libraries, health care institutions, telecentres, and other community centres, including those in remote Aboriginal communities.

This more incrementalist infrastructure approach is a compromise in terms of planning for ubiquitous and affordable access to new communication services. It will be fascinating to see how many initiatives arise from the Howard government proposal to take USOs to tendering. Strategically, Australia has no choice other than to build an advanced networked society for the future. There are issues and alternatives about ways and means, but little doubt about the desirable goal. We need to integrate the construction of the new communication platforms with understanding and promoting the demand for information and communication services which meet people's needs. Getting the new communications order right for Australia—thinking global business/community opportunities and benefits—is third way thinking indeed. This will be vexed and require substantial investment in an intensely competitive global world. The economic process may not be particularly satisfying—rather like rowing upstream against a tide which seems to be getting stronger, only to acknowledge that at least our position has been maintained, or that at the very least we are still in the water.

Digital déjà vu

Another vexed issue concerning the development of a third communications order in Australia is how convergence policy unravels in the next decade. There are some complex tensions between the established broadcasters and the prospective new on-line service providers. While the spirit of competition policy has been to open up the marketplace—and the spectrum space—to as many different players as possible, this had been politically difficult in practice. The policy debates about the introduction of digital television to Australia in the late 1990s saw a renewed phase of jockeying for new spoils among the media power brokers. This involves a technological generational change into digital technology, and in the late 1990s the established commercial television interests again manoeuvred successfully to preserve and enhance their patch. As Terry Cutler put it, '[T]he way media policy is set has all the subtleties of Mafia gang warfare, and the case of digital television is no exception. Self interest is so brazen that it goes totally unremarked.'[20]

Digital television uses advanced digital techniques to convert an analogue signal, along with other signals, to a compressed digital signal before being broadcast from a transmitter. Essentially, it can provide high-quality images to television sets together with greater broadcast capacity and signal diversification. Digital terrestrial television broadcasting (DTTB) can:

- provide for better reception of television services than the current analogue systems, including the PAL system used in Australia;
- deliver higher quality picture and sound than the PAL system;
- provide high definition television (HDTV) programs or a number of standard definition television (SDTV) programs within the standard broadcast channel;
- reconfigure services (that is, change from a single HDTV program to a number of SDTV programs) at any time; and
- carry a range of multimedia services in the form of audio, images, data and text.[21]

What it means for Australian consumers is that they will need to buy a digital set-top converter box at an estimated cost of about $400. Alternatively, they might choose to buy a new integrated high definition television set for about $6000, although equipment prices ought to go down as take-up goes up.

The introduction of digital television in Australia involves vexed decisions about the allocation of spectrum space, substantial capital investment by the television licensees in both the public and private sectors, the prospect of opening up the market to new players, the possible diversification of services, and the purchase of new television sets by consumers—all the ingredients for renewed struggles about power and influence in Australia's lucrative over-the-air television marketplace. Inevitably, this was going to be a tough call for government in the context of new media services and assessing the thicket of vested interests, old and new. Briefly, the positioning by stakeholders in the lead-up to the policy decisions may be summarised as follows.

Commercial television interests argued that this was 'a quality issue, and one that is being forced on television by its competitors' in that digital television transmits a far better picture than analogue systems.[22] Commercial television argued that they needed 7 MHz of spectrum space to broadcast high definition television and flexibility in the mix of digital services they provide. And, they added in this age of competition and deregulation, 'no new commercial licences should be issued during the transition period. This is necessary to provide the minimum degree of certainty required for such large investment by broadcasters.'[23] Everyone else was to face more competition in their businesses, but not them.

The ABC argued that 'digital television puts a new means in our grasp', called for a full allocation of 7 MHz of digital spectrum, and proposed that the ABC could raise more than half the $110 million it needed for the first stage of digitalisation with supplementary funding to come from government. Managing Director Brian Johns argued that, otherwise 'the ABC will be unable to properly fulfil its role as national broadcaster'.[24]

Telstra warned the Minister that several of its commercial interests in data and on-line services could be affected if the free-to-air broadcasters were awarded digital spectrum at no cost. If the free-to-air broadcasters wanted to move into data and on-line services, argued Telstra, then the spectrum should be allocated under the *Radio Communications Act*, which would involve a competitive bidding process, rather than as a giveaway under the *Broadcast Services Act*.[25]

The Australian Subscription Television and Radio Association (ASTRA), on behalf of pay television and other subscription service industries, argued against 'the last available spectrum space given up in the pursuit of television interests of network television, with no consideration given to the exploding digital communications world—Internet, Internet telephony, on-line applications, narrowcasting, datacasting and interactive services.'[26] ASTRA proposed that the existing commercial television broadcasters should be restricted to the transmission of a single digital standard television service.

A coalition of commercial and consumer groups, including News Corporation and John Fairfax, no longer with television licences in Australia, argued that the networks should not be given an unfair advantage in controlling the 'gateway' to the next generation of digital services.[27]

Consumer groups argued that it might be best to work out what the spectrum should be used for, then maximise its use.

After a great deal of lobbying cut and thrust by the major vested interests in convergence, the government announced several major decisions late in March 1998:

- HDTV will begin on 1 January 2001 in metropolitan areas and progressively from 1 January 2001 in all regional areas, so that all areas have services by 2004.

- The existing free-to-air TV networks get free spectrum for digital broadcasting until 2006 with no new free-to-air competition until 2008.
- Free-to-air broadcasters cannot use the capacity for multi-channel digital services until 2005 (the main fear of the pay TV operators). A review would determine whether the ABC and SBS might be able to do so where programs are 'non-commercial' and 'in line with Charter obligations'.
- Existing networks are allowed to use digital for 'enhanced services', which might be information services related to their television programs, or for 'datacasting'—a hybrid of text, video, speech, images and other forms—which may be accessed in future through the television set. The ABC and SBS can sub-lease capacity for commercial datacasting, subject to revenue sharing with the Commonwealth. Subsequently, the Australian Broadcasting Authority recommended that each free-to-air station be given an extra channel on which to broadcast its digital service; Channel 7 should broadcast digitally on Channel 6, Channel 9 should have Channel 8, Channel 10 should get Channel 11, the ABC Channel 12, and SBS Channel 29.

Clearly this was another great policy victory for established free-to-air television interests. Only commercial network television interests, it seems, can get long-term policy commitments out of governments. These decisions were met with howls of protest from the daily press, now owned by the Murdoch and Fairfax interests, who no longer have major television interests in Australia. Murdoch's *Australian* editorialised 'Government plays favourites with TV' (26 March 1998), and Fairfax's *Sydney Morning Herald* ran with 'Digital TV giveaway' (27 March 1998). Meanwhile, the *Australian Financial Review* (25 March 1998) editorialised that 'Senator Alston has simply shackled the new information economy in the familiar old world of heavy government regulation and media-mogul politics', and its Chanticleer columnist, Ivor Rees, described the Minister on the same day as 'the ultimate political pragmatist and the media moguls' best friend'. Not a good press for the Minister, the government and the commercial television bosses to begin their digital days.

The digital television policy decisions have deeper significance than merely *déjà vu* of the days of lucrative television licence handouts to media moguls. For better or worse, all sectors of the Australian economy have now been forced to embrace national competition policy. Not, however, the Australian commercial television free-to-air industry in terms of direct new players in their core business. Leaked Cabinet documentation subsequently published in the *Australian Financial Review* (7 May 1998) revealed that the government ignored high-level advice that this was an unwarranted 'free gift' to the commercial television networks. The Department of the Prime Minister and Cabinet had argued that 'the proposed approach would restrict the consumer benefits available from a wider range and flexibility of services and could have significant budgetary implications through forgone revenue from the sale of spectrum'.[28] Moreover, if we are serious about building an information society, we need to grow the services of convergence and devise policies which allow a plethora of new players into the new order. The Department of Finance said that the plan 'was inconsistent with the reforms which the government has made in relation to telecommunications industry regulation and the non-interventionist approach which the government has endorsed for developing the information economy'.[29] So, one arm of government policy claims that we must make 'technology-neutral' decisions, let things rip, and leave it all to market forces in the new on-line culture; while the other arm of policy was making decisions which provided for the long-term protection of status quo non-competitive vested interests.

Also at a time when government has moved towards the auctioning of spectrum space, only one group has been quarantined from paying for their new spectrum rights. This decision enhanced privileged access to public assets, and revitalised and revalued commercial television licences. Moreover, the networks argued that the move into HDTV will be expensive for them and consequently these investments are to be cross-subsidised by other taxpayers. Yet, other major Australian corporations who have to invest in new generational technologies, such as mining and agricultural corporations, retail giants and banks, information technology or telecommunications companies, have to organise such recapitalisation themselves *without* government assistance. These organisations, however, do not run national consciousness outlets in election years.

It is incredible in times of the most rapid technological and service changes ever in communications, and when international policy is so driven by competition policy, Australian government policy in 1998 could lock one group into spectrum space without direct competition in their core business—until 2008! It is most likely that the Murdoch camp will put great pressure on this decision. Indeed, Executive Chairman Lachlan Murdoch announced in November 1998 the creation of a new company, News Broadcasting Australia, to develop services in areas as diverse as datacasting, narrowcasting, multi-channelling, education and e-Commerce, and potential pay-per-view-systems. Asked if the new company was created to seek a fourth commercial television licence, its new head, Jim Bloomfield, said: 'Yes, it is for that . . . No other industry has been protected the way this has been.'[30]

Surely there is a case to keep policy options open in terms of stakeholders at a time when the technological choices are so fluid and dynamic. As Michael Ward of OzEmail put it at a Sydney seminar in February 1998, 'The question that we need to consider in allocating bandwidth for fifteen years is whether somebody will come along in the next month or two, year or two, ten years or so, with a better idea for how to use some of that bandwidth. There are already people who have better ideas about how to use that bandwidth.'

Ward's judgment is shared by plenty of others, and irrespective of whether comparable bandwidth is available but remains unused, surely a major communications development policy direction is towards open systems, open access? Why did policy-makers shut out new options for significant development opportunities in communications services and practices? This was a government that came to office on the promise of promoting competition everywhere and making 'technology-neutral' decisions! A bi-polar communications development strategy based around traditional broadcasting on the one hand, and the new interactive datacasting services on the other, does not make sense.

A unique opportunity to develop an imaginative digital convergence policy has been lost.

9 Re-thinking our communications strategy

> *It's time somebody stood up for economic rationalism in the midst of all this stupidity. Economic rationalists are concerned about efficiency and common sense . . . by wherever possible maximising the opportunities for markets, market forces, market disciplines and market processes to operate.*
>
> John Hewson, former Leader of the Opposition, in 'A classic piece of irrationality', *Australian Financial Review*, 24 July 1998, p. 32

> *If we are to survive and prosper in the third millennium, we need a revolution in the way we think about and pursue personal and social well being . . . The primary function of government in a post-growth society should be to protect, expand and enrich our human, social and cultural and natural capital.*
>
> Clive Hamilton, Executive Director, The Australia Institute, 'The dark side of the Australian dream', on 'Ockham's Razor', ABC Radio, 20 June 1999

Information and communications are now so intertwined economically and socially that they are part of the fabric of the way we live our lives. Information and communications systems and processes do not sit outside the mainstream economy, neatly compartmentalised as a 'nice' or 'useful' economic area to develop. Debates about our economic development have generally been couched in terms of how we might develop particular sectors of the economy, such as the steel, automotive, textiles, wine, tourism, education, gaming and gambling industries. These sectors, however, have all now become more dependent on communications to operate more efficiently and to expand globally. It is now difficult for any business to grow without access to the range of new communications services. Convergence has also generated multiplier effects across the economy. For instance, every time someone dials up an Internet service provider, a telecommunications carrier gets the cost of a local call; every time the ISP connects the subscriber to the Internet, the subscriber pays them; the ISP in turn may pay the carrier an interconnection fee—but the carriers get the bulk of the revenue derived from the data traffic. *IT&T has become the enabling technology of the late twentieth century.* Converged communications activities and practices are now generic to development because they permeate virtually every aspect of economic and social life in Australia.

It is essential that Australia builds its own advanced information society for the future. We urgently need to come to terms with the links between globalisation and its interdependence with information technologies, the structural changes occurring within modern contemporary economies, and the centrality of new communications processes to economic prosperity and social well-being. The significance of the development of an advanced information society in Australia has been subject only to limited general public debate and given scant attention by the major political parties. While there is no absence of views as to how we might develop certain aspects of our communications industries and practices, and plenty of vested interests pushing their own barrows, there has not been any thorough and systematic canvassing of national strategic issues, or any real sense of the emergence of integrated policies. We need to decide what kind of a communications environment we want for Australia and aspire towards it. *The centrepiece of public policy during the 1990s—belief in economic notions of efficiency and competition—is too narrow a policy framework for Australia in the future.* Of course, Australia must become more efficient and competitive, but these ought to be means to an end rather than an end in themselves. We ought to spell out our goals in some detail—in terms of industry employment and export opportunities, diverse communication and cultural services that are needed, widespread access to new infrastructure, local development prospects for small and medium-sized Australian companies, and fostering innovation in new media—and use the appropriate policy tools to achieve our objectives. We need to realise far greater potential in the relationships between contemporary communications growth and widespread community benefits.

Politics, not policies

Technology and communications policy in Australia has increasingly been characterised by an appalling neglect of social philosophy. We have had little sense of the need for systematic investigation of options for the future, of integrating good economic and social planning, and of building a strong community awareness of the issues surrounding participation in communications policy-making. The ad hoc decision-making in communications policy is the inevitable consequence of an essentially irrational Australian political process from which clear-sighted integrated policy statements have failed to emerge as blueprints for action. A small, elite group of policy-makers holds power in the key decision-making process. This group is composed of a few transitory politicians who are members of the elected government of the day, together with their key advisers, who call upon the more permanent senior members of the bureaucracy, with selective inputs from certain commercial interests and politically effective pressure groups.

There is no permanently established mechanism in Australia whereby a wide and influential cross-section of the public can effectively be informed of the issues and offer an input into the policy decision-making process, analyse the policy outcomes and constructively propose change. It is little wonder that we have had a series of isolated and fragmented communication policy reports, which, although they have offered some valuable insights into parts of communications policy agendas, collectively represent a serious failure to devise an integrated, workable national strategy for Australia's communications future. The serious flaws in the public policy process do not constitute a case for its abandonment. The alternative is to leave everything to the vagaries of the marketplace—and to live with the inevitable serious consequences of the growing polarisation in Australian society.

Barry Jones has offered some salutary reflections on his long involvement in Australian public policy. His book *Sleepers Wake! Technology and the Future of Work*, first published in 1982, called upon Australians to understand that information-related work would become by far the largest employment area with the greatest potential for wealth creation. He has long argued the need for a National Information Policy, but pointed out that 'the House of Representatives, in the nearly twenty one years I was a member, never had a serious debate about the scope of Information Policy, irrespective of which Party was in power. Neither did the Senate.'[1] He further points to the territorial approach of the Commonwealth bureaucracy where the departments each interpreted 'information' differently—the Department of Industry equated 'information' with IT, the hardware; the Attorneys-General with intellectual property, patents, copyright and privacy issues; the Department of Communications with telephones, radio, television and program content; the Treasury with statistics; Education with education and training; and Arts with the National Library, film and arts funding. For Barry Jones, the policy process, in what he referred to as Information Policy, was highly fragmented and lacking in any sense of cohesion and national direction.

Underpinning these criticisms lies a fundamental power shift in contemporary Australia—the deconstruction of the role of the state in ownership, policy and planning for the future. In recent years this has manifested itself in many ways in communications policy:

- the partial privatisation of our national telecommunications carrier, Telstra, which will inevitably lead to its full privatisation;
- the abandonment of industry development policies in favour of generic competition policy;
- our failure to use government purchasing policies to foster the growth of small and medium-sized Australian manufacturing companies;
- the shift in the prime role of regulation from notions of serving the 'public interest' to monitoring 'structural regulation'—the purpose of which is to facilitate unfettered market-based decision-making;
- the replacement of communications infrastructure planning in the public sector with an assumption that a multiplicity of new private sector players involved in cut-throat competition will somehow be able to invest in long-term infrastructure projects for the communication networks needed for the future;
- the savage cuts in the budget allocation of the public broadcasting sector and the deliberate destabilisation of these public institutions, notably the Australian Broadcasting Corporation; and
- the abandonment of any political vision associated with public funding for research and development grants in information technology; of attractive incentives for investors to support our film industry; and of real measures to foster Australian drama, Australian children's television, and experimental film and television forms for young Australians.

We have lost our way with innovation policy.

Why has this happened? There is obviously no simple answer, although we might do well to revisit the work of the German philosopher Jürgen Habermas who posed many questions about technological developments and the institutional framework of post-industrial society.[2] For Habermas, a new kind of technocratic society has emerged in which technocracy 'signifies a social order organised on principles established by technical experts'.[3] We now live in an age of technocratically managed capitalism. Politics has, in his view, become basically reduced to a question of who can run the economy best—perceived as merely a matter of technical decision-making. Modern capitalism has seen the emergence of pragmatic, technocratic government, where the common people have become alienated from a political system which has failed to maintain its own narrow self-interested objective of sustained economic growth. For Habermas, a fundamental problem arises because the social system does not obey the logic of scientific progress—values and interests clash, social choices arise and complex questions need to be reflected upon. It is within this process of re-evaluation and reflection of a new world paradigm that the communications order has special responsibilities and opportunities for citizens.

Habermas argued that an autonomous and vigorous public sphere is being undermined in contemporary capitalistic societies, now enmeshed within the new global economy, so that the political decision-making processes have progressively abandoned public scrutiny and collective will formation. For Habermas, 'a legitimation crisis' has emerged in contemporary society because former public accountability, for so long the hallmark of civil society, has been undermined and progressively abandoned. The critical theory of Habermas deals essentially with the relegation of the representative structures of the formal democratic processes to an intermediate role in society. Do we want to accept that we, as Australian citizens, can do nothing about this legitimation crisis?

Discourses for development

So, what do we do as a nation about Australia's new media for the new millennium? There is no obvious and clear plan to guide the future of Australia's communications, but there are plenty of vested interests pulling in different directions. *There is a lack of a common discourse for the development of national communications policy for the future.* Key stakeholders approach the complex issues of what should be done with communications from different positions, with varied interests and agendas, usually with little understanding of other positions. Corporations want to maximise their profits and achieve good annual financial returns for their shareholders. Engineers want to build the finest technical networks. Economists want to impose their sense of an efficient economic model. Lawyers want to reduce the risks for their clients. Academics want to see more systematic research to underpin decision-making, in both the public and private sectors. Consumers want cheaper prices and better services to be delivered to them. Citizens want assurances that the new communications services will be accessible and readily affordable, if they choose to use them. Hence, the lack of any sense of commonality among these disparate groups creates a crisis of meaningful discourse. These clashes of interest are manifest in the variant languages used by different stakeholders in communications policy debates. In most communication policy forums, everyone, it seems, is talking at cross-purposes.

This book has analysed the new Australian communications environment in the context of many complex forces for change. Three key themes have been central to the arguments presented throughout this book.

First, the world's economies are increasingly interdependent and on an unprecedented global scale, facilitated by the extraordinary technological sophistication of new communications networks of data and information. No government can ignore the economic consequences of its political actions, lest the financial markets of the 'global economic casino' slash the value of their currency, possibly tumbling their economy into serious debt and driving away potential foreign investment. In this context, Australia's economic policy will essentially be accredited and governed globally. However, *although nation states may have lost a significant degree of their economic power through globalisation, they have not lost their influence; economic globalisation may be inevitable, but cultural globalisation is not. Australia can still fashion critical aspects of its own destiny.* There is likely to be a shift away from the extraordinary power of the global currency markets in the next decade, with more regional and national attempts to halt the alarming flights of capital resulting from the ruthless daily anecdotal assessment of national economies.

Second, a nation's communications order—the communications infrastructure and nature of its communications services and processes, and how national communications policy is constructed—will increasingly determine its level of prosperity, social well-being and national identity. *The kind of society Australia attains will be dependent upon the character of the new communications order it constructs.*

Third, the benefits of the information revolution are not dispersed in a planetary, global or national way; rather, there are highly selective winners and increasing divides between the information rich and the information poor, collectively for societies, and individually for citizens. *Australia must construct national policies which allow the benefits of the communications revolution to be harnessed by all Australians who want to participate.*

We need to embark upon fostering greater community awareness about the complex issues of the construction of our national communications policy. Communications has become our favourite growth business, yet we have little collective imagination about its national benefit for Australia. We have few imaginative policy thinkers. We need to foster attempts at constructing national communications plans for Australia's communications future.

Towards an Australian communications policy blueprint

1. *Australia must continue to foster the development of communications as central to building an outward-looking, internationally competitive productive economy.*

The Australian economy has changed structurally in recent years, away from its overdependence on agriculture and mining and towards a necessary broadening of its export base, but it still remains vulnerable as essentially a commodities-based economy. We must add value to what we do. Australia must nurture and develop the culture of converged communications to our national benefit in this new global society. This ought to involve *all* of our institutions, both private and public, in long-term strategic planning. Those who flinch at this on the grounds that governments ought not to embark upon a strategy of 'picking winners', or who warn against the undesirability of pushing particular forms of sectoral assistance, do not understand the terms of these debates. The notion of a converged communications economy *is* the biggest 'winner' in contemporary economies around the world. The strategic thinking and level of decision-making required must be at a higher level than merely tinkering about, whether we support a new invention or product, or particular policy modifications such as what are the desirable tariff levels, or whether to provide government assistance programs, or disputes about taxation reform. The wealth and good social potential of contemporary communications are not in question. The questions for Australians are: how well do we measure up in the new international economic and social communications order, where do our opportunities lie and how might we position ourselves for the future?

Although parallels between the Australian and American economies cannot be easily made, we might do well to emulate aspects of what is referred to now as 'the new economy' of America in the 1990s: an economy built around contemporary communications. The American economy has been characterised for more than a decade by steady annual growth of its GDP: 18 million new jobs have been created since 1992, unemployment is at a 30-year low, and inflation has been kept in check. *The Economist* argues that 'a new economic paradigm is sweeping America . . . computers and globalisation have brought faster productivity growth that permits rising profits, rising wages, and falling inflation, all at the same time'.[4] A University of Texas study argued that the Internet was responsible for generating 1.2 million jobs and pumped US$301.2 million into the American economy in 1998. One in five workers employed in the US technology sector now have Internet-related jobs.[5] Even if there is a substantial share market downturn in the United States in the future, the vital characteristics of their value-added economy will remain in place. The jury is still out, however, on whether the benefits of America's economic dynamism will lead to richer lives for most of its people, or whether America will be a better place in which to live.

Australia has to nurture communications as a part of a necessary economic transformation to secure our prosperity for the future.

2. *We need to revitalise the notion of the public sphere.*

In the late twentieth century the role of the state is fundamentally in question: communism has collapsed and capitalism has gone global. Ironically, a time of unprecedented national communications policy planning by governments around the world is coinciding with major withdrawals on the part of government from the business of industry. The politics of debt during the 1980s and 1990s has enabled economic rationalists to capture the policy process in Western democracies. The slogan is now widely accepted that 'best government is the least'. Both the established role of government, and the major contribution made by public institutions, are under challenge. Communications has a proud history of major technological innovation that emanated from the public sector; the development of communications satellites as integral to the American space program in the 1960s, and the foresight of the American defence bureaucracy that conceptualised and developed the early stages of the Internet, are notable examples. Whereas at the beginning of the twentieth century telecommunications was considered too important to be left to the *private* sector, at the end of the century telecommunications has come to be regarded as too important to be left to the *public* sector. Moreover, in broadcasting, the valuable contribution made by public broadcasters to democratic expression and cultural development is increasingly being questioned. Balances of power between the public and private sectors have shifted towards a stronger market-based ideology, with extensive privatisation, deregulation and liberalisation. Will this make Australia a better society?

We have witnessed a power shift in Australian society during the past decade or so towards a more private, corporate, deregulated economic paradigm. This power shift has led to the persistent calling into question of the role and legitimacy of the public sector and created an identity crisis for public institutions. Our media and communications system has long been built around a mixed economy of public and private sectors, but with significant public institutions, and government intervention in certain practices, including by regulation, has been defended on the grounds of being 'in the public interest'. Both the public and private sectors have long played important and complementary roles in our society. Why can't we achieve better integration between the public and private sectors for the purpose of planning development nationally? Public policy-makers, public broadcasters and public telecommunications carriers can adjust to new times and still maintain the best of their traditional roles and values in protecting good citizenship.

The public sphere signifies the arena of conflict and contradiction in which particular opinions, positions and aspirations clash with each other, but expression and distillation of the wide range of discourses is critical to effective democratic decision-making. Surely it is better in terms of the national interest to work through these differences and divisions, no matter how difficult this might be, than merely to allow the most powerful vested interests to have their own way. The construction of effective public policy for the future requires that the complexity of discourses be addressed. It is the task of elected governments, Commonwealth and state, to come to terms with these discourses, to work through the multiple policy options, and eventually to take the tough decisions that are in the best national interest. This most fundamental role of the state has not changed.

Do we really want to leave virtually everything to whoever rules the marketplace? Most Australians feel disenfranchised from the major political processes that affect their future. We need to ensure that the complexities of contemporary communications policy are more widely canvassed and debated, and that there is a greater awareness of the alternatives. The notion of the 'third way' political order means more than building our economy to be globally competitive, but the third way also calls for the renewal of social democracy. The continuing deconstruction of the public sphere can only result in a more disenfranchised and disillusioned Australian citizenry in the future.

3. *We need to broaden our sense of the meaning of development into long-term, imaginative national strategic programs for future communications.*

The decade of the 1990s has been characterised by an aridity of thinking about how we see opportunities for development. Competition has become an end in itself, rather than a means to an end. The corporate sector has become ruthlessly utilitarian about its short-term financial performance, policy-makers and regulators have been captured by narrow thinking about economic efficiency and competition, and public sector institutions have found themselves under siege. There appears to be select rewards for significant economic growth in particular sectors, especially communications.

We need to broaden the base of our decision-making about the future of Australian communications. More people are needed from different backgrounds to become involved in our thinking and planning. What is required is strategic thinking about our future, rather than narrowly based strategic development dominated by industry groups. The creation of multiple-ideas communications blueprints needs to be fostered. We need more plans to be devised by the 'thinkers' in communications, involving members of both the public and private sectors and with good input from the broad citizenry. This process should emerge from the public sphere and must not be devised or perceived as a 'top down' Canberra-based interventionist program. The creation of a 'communications think tank' will need to be 'sold' politically to governments and political parties to get them to understand how contemporary communications has become central to our future national well-being.

Part of this strategy would involve finding constructive ways of using the communications industry to contribute to addressing Australia's unacceptably high levels of unemployment and underemployment. The history of the communications industry in Australia is that it was long regarded as a vehicle to create employment opportunities. The PMG, later Telecom, progressively rolled out its network, a local electronics industry was built around it, networks were progressively upgraded and important service industries emerged. Yet in the 1990s the employment-generating policy for our leading carrier was reversed—Telstra's redundancy program involves the loss of 27 500 jobs by 2001 and the corporation is 'well on track' to reach this target. Many of the functions of Telstra (and other large Australian corporations) are now being outsourced, and only time will tell whether many important company activities run more efficiently on this external itinerant human resource model. Will there always be an inner core of key staff who know the business well enough to commission and manage the outsourced?

New network structures built around information economies do provide many new economic opportunities, new ways of communicating with each other and new forms of human creativity, but they are also contributing to greater polarisation and isolation in society. Capitalism has always produced both winners and losers, and we are accentuating the serious income divisions and widening class divisions between the information rich and the information poor. Equity issues must be addressed by national policy-makers in our new economic order of the user-pay society in Australian society. We must address our alarming levels of unemployment and foster new work opportunities in communications. We need to build a value-added economy with associated satisfying work opportunities, and not just accept that many of the new jobs in communications will continue to involve dull, routine work, possibly in call centres or for isolated teleworkers.

Our present education system possesses more talented students than ever before in the history of the nation, yet too few students now have real opportunities to find satisfying avenues of work. The sad paradox is that a booming Australian communications industry provides few new satisfying work opportunities for all sections of our community. The middle level of employment opportunities, which historically provided fulfilling career opportunities, is diminishing. Do we really expect that the present policies, founded on neoclassical economic assumptions about growth, will somehow eventually provide meaningful work for more people?

Is the long-term acceptance of jobless growth efficient for the nation? Why do we think of human development programs as a cost, rather than as an investment for the future? Is it efficient to spend heavily on an education sector only to outlay *additional* public expenditure at the end of the education process on work for the dole programs? What is the human toll in this? How do economists measure that?

Public national communications development programs would be readily affordable. In the field of communications, Australian governments are rewarded by what might be called 'double dividends' from certain practices that raise substantial revenue from various financial calls made on communications companies and organisations by the Commonwealth of Australia. This income for the state is additional to the revenues raised from normal taxation collection of media, information technology and telecommunications companies. In the 1990s the field of communications has been a goldmine for the Commonwealth government. A licence fee of $850 million was paid in 1991 by the new telecommunications carrier, Optus Communications, to compete with Telstra as a duopolist fixed carrier until 1997. There are also significant government dividends paid by publicly owned companies, such as the extraordinary dividend payments of $2–3 billion *per annum* by Telstra in the second half of the 1990s. Substantial licence fees, amounting to almost $200 million in some cases, were also paid by particular companies who bid for pay television licences. Licence fees are paid by all commercial radio and television companies to the Commonwealth, to the order of $200 million per annum. Another continuing source of additional communications revenue for the Commonwealth is the fees raised by the Spectrum Management Agency. *Collectively over the decade of the 1990s, the Commonwealth government appears to have raised around $20 billion from its 'double dividends' from the communications sector.* No other industry has had to be so generous to its owner or master!

In the last decade the Commonwealth has been flooded with cash from this sector. Yet the Howard government in its 1996 budget reduced the R&D tax concession from 150 per cent to 125 per cent, which has resulted in an alarming downturn in business R&D spending. Why doesn't Australia follow the rest of the developed world, where a 200 per cent R&D tax concession is common? Can't we afford to heavily invest in R&D as a nation for our future? Why aren't we providing more seeding money for incubators to foster the talent in small start-up companies with high growth potential? Why aren't we thinking strategically about ambitious national communications development strategies? This is not a case for propping up inefficient sectors of the economy, or proposing random handouts designed by bureaucrats, but for embarking upon important strategic investments of many kinds—such as new advanced infrastructure, the promotion of on-line innovation, content initiatives, youth participation in new media projects and the promotion of national Internet literacy. Why not invest now in productive communications programs for the future?

4. *A vital component of the construction of an Australian advanced networked society is to make affordable high-bandwidth telecommunications available, preferably to all homes, institutions and offices.*

We might well remember that the telecommunications infrastructure we built during this century is the nation's most valuable physical asset and that we are in a strong position to further capitalise on the network society that has emerged around it. We have considerable development capability in the communications systems and practices which have been critical to the growth of contemporary capitalistic economies. For a long time Australia had an unswerving public policy commitment to building a national communications network. Until the mid-1960s, this was directly cross-subsidised by taxpayers, as well as from the public carriers' internal tariff policy for inner metropolitan telephone subscribers. This policy commitment broke all the rules of modern-day textbook economic efficiency—it was driven by government policy, it was 'trying to pick winners' (which it did brilliantly), it meant that taxpayers as a whole were paying for the huge costs of this development, and it inflated and skewed telecommunications pricing for many users while heavily subsidising others. The process was managed by a public utility which became the corporation with the greatest market capitalisation in Australian corporate history, reaching over $100 billion early in 1999—Telstra. Posterity can only judge favourably such long-term strategic investment in this public infrastructure program. Yesterday's telecommunications policy-makers had an intuitive strategic sense of long-term benefit; today's economic rationalists would never have built this superb nationwide network.

In the era of competition, telecommunications carriers are now reluctant to become involved with national universal service obligations. Why would a carrier want to contribute to funding new infrastructure in the bush when it can make a much more financially attractive investment in, for instance, high-speed, inner-city Internet services? As we saw in Chapter 8, the policy-makers have walked away from a fundamental social policy in Australian telecommunications—universal service at affordable cost. Yet why have a policy of promoting companies to invest heavily in new on-line services but then not ensure that consumers have the network to access those services?

New communications services offer the real prospect of assisting people who feel somewhat dispossessed from society—for example, members of some rural communities who have lost their local school, local bank, local railway and local post office on the grounds of the lack of viability of these distant services in the 1990s. Some have also lost their sense of place and belonging in contemporary society. In a small but important way, rural Internet cafes and learning to tap into the Net can provide a renewed sense of connectedness. Better, of course, if we could provide them with access to the Internet at comparable levels to telephony access. Many of the big department stores may be empty in South-East Asia, but there are plenty of people lining up in the arcades to get on the Net for a short time. In these senses the Internet has made a positive contribution to the notion of social renewal with the death of distance.

We must maintain the USO fund, but extend it conceptually. In essence, what we need to do is to maintain the equity principle of this fund and use it to upgrade our national networks. We now need widespread access to more than just telephony, and we must work towards building networks which provide high-bandwidth telecommunications services, preferably for everyone. We must recapture the sense of vision that understood for nine decades of the twentieth century the need for long-term major investments in national strategic infrastructure. The new networks, however, will be built by a multiplicity of carriers, using a hybrid of delivery technologies where satellites and network software will be more important than ever before. We can build the new networks for the future in stages, beginning with funding public access points as universal reach and work progressively towards universal access. We need to think in terms of universal service opportunities, rather than merely universal service obligations.

5. *We need to think of our new networks as integrating new business and community purposes. E-Commerce is the new beachhead to secure our national trading prosperity, and e-communities offer the diversity of global communication processes that 'old' forms of centralised institutional media never provided its citizens.*

The Internet may become the platform for the most radical way we do business in this new century. Upgraded networks are a basis for the flowering of new types of businesses now emerging around the Internet. The development of electronic commerce requires a considerable change in the corporate ethos of the Australian private sector, which has generally been slow to comprehend the extraordinary changes that information technology has facilitated. There are plenty of businesses that are finding that new modes of marketing and distribution through electronic commerce, now a $10.5 billion business in the United States and worth $85 million in Australia, are progressively undermining their traditional practices.[6] Does a traditional bookstore simply ignore the world's biggest bookseller, amazon.com, which sells a million books a day on the Net, or does it try to compete on the same terms or in some other way? There are some tough decisions here for Australian companies and institutions, and the signs indicate that we are slow at positioning ourselves. An Andersen Consulting report showed that while 80 per cent of CEO respondents agreed that e-Commerce will revolutionise the way they do business in five years, 81 per cent of them did not include e-Commerce in their top four strategic priorities.[7] If Kevin Kelly, distinguished editor of *Wired* magazine, is right in predicting that 20 per cent of world trade will be via electronic commerce by 2003, perhaps we urgently need to elevate the priority given to electronic commerce in the boardrooms of Australian corporations.[8]

The Australian e-Commerce business cases highlighted in Chapter 5 are surely indicative of a classic latent market. Australia does have a culture of creativity, a history of rapid technology uptake, including one of the highest Internet participation rates anywhere, world-class telecommunications networks, and ample technological 'know how' to build successful e-Commerce businesses. In terms of national productivity in a globalised world, there may be few choices here. The world may be embarking on an era of new retailing imperialism through electronic commerce, and the United States is highly likely initially to gain the greatest benefit from this new mode of merchandising delivery. The risks of non-participation for Australia could be a seriously worsening long-term trade deficit.

There is little point investing in the new infrastructure for the e-Commerce/e-communities order unless we can nurture, foster and encourage substantial numbers of users of the new communications highways, and provide them with a sense of confidence in these new modes. How do we build this? Some innovative Australian workgroups convened by CIRCIT have explored possible actions which might enable the greater utilisation of on-line services. Their suggestions may be summarised as follows:

- identify 'icons' or 'champions' who can promote the benefits of on-line services to the broad community;
- promote user skills and training across all sectors;
- recognise the leadership vacuum in corporate Australia with the agenda being driven by technical experts rather than senior operational and strategic business executives/board members, many of whom are not connected to the on-line world;
- create new on-line business incubators;
- continue to support and promote applications developments, including trials and evaluation . . . There are applications going to market . . . with no apparent ongoing evaluation process built in . . . we need to have impact assessments/statements in terms of user issues;
- identify leading-edge projects which can be used to highlight effective use;
- target young content developers and provide support to develop concepts; and
- provide incentives such as reductions in charges provided by the New Brunswick (Canada) government for on-line transactions.[9]

To promote the development of an on-line economy we could broaden our concept of R&D tax breaks to include service companies.

This is also an area with great employment-generation potential, particularly for young people who have a real feel for the new media. Why not support such ventures above with a highly affordable 200 per cent R&D tax concession for new media service companies with talent?

On-line innovation also has great potential for Australian software developers. As Simon Molloy has pointed out, '[T]he Web demonstrates the fundamental economic truth of extensive networking: common standards over large domains create new markets, innovation and growth.'[10] The days when computing power was based on expensive proprietary and incompatible networking standards are over in the Internet age of any to any connectivity. There are several main reasons why Australia ought to be able to do well in the new software revolution: the relatively lower development costs; the 'death of distance', which ought to mean that anyone anywhere in the world can participate; and an education system which produces first-class software engineers. Although the promise is there, it remains the case that the dynamism of the US capital markets, and the technological leadership and global scale of the Microsofts et al., mean that the United States has a commanding position. However, as the on-line economy becomes increasingly customised to meet particular users' needs in particular environments, the opportunities ought to blossom for Australian on-line software developers. Both public and private sector policy ought to support such innovation, and be prepared to invest and to take risks.

The metaphor of the Web suggests the unique connectedness of e-communications in global 'any to any connectivity'.[11] The Net offers us an invaluable research tool, enables us to purchase goods and services if we wish from many organisations and places in the world, to communicate via e-mail with friends and associates scattered around the globe, and to form common interest groups anywhere with virtually anyone. The old established information institutional gatekeepers are being progressively undermined, and there is the potential for the revitalisation of the meaning of citizenry.

6. *We need to create a new convergent communications regulatory body to monitor and promote the new industry dynamics and opportunities for Australia in communications.*

Convergence needs new rules for new times. Traditionally, media, information technology and telecommunications were separate, each offering distinct services, such as broadcasting, voice telephony and stand-alone computer services. They also used different platforms, such as television sets, telephones and computers, and essentially used different networks to deliver their services. Each was regulated by different laws and to different degrees, with much less regulation in information technology; broadcasting and telecommunications were generally accountable to different regulators established by national governments. However, in the world of new communications, with digitalisation and high-capacity networks, regulation is complicated because different services are transported over the same networks to provide integrated consumer offerings, such as the emerging on-line services. The lines of demarcation are becoming more complicated due to convergence.

Media, IT and telecommunications companies are using the flexibility of digital technologies to diversify their traditional services and embark upon these new markets of convergence. In summary:

- Broadcasters are exploring new areas, such as data broadcast, Internet webcasting and telecommunications services.
- Telecommunications carriers, with a history of service in voice telephony, are now providing a range of audiovisual services, such as subscription television and video on demand, as well as new data services, such as for financial markets and multiple services via the Internet.
- Internet service providers are also beginning to distribute audiovisual content, and Internet access providers are beginning to offer voice telephony.

In this blossoming new information services marketplace, examples of new convergent services include:

- newspapers with web sites—3500 newspapers in the world now have Web versions of their newspaper;
- television channels offering web sites associated with their own programming;
- home banking and home shopping over the Internet;
- Internet services delivered to television sets, via systems such as Web TV;
- webcasting of radio and television programming on the Internet;
- voice telephony using the Internet; and
- e-mail and Web access via digital television decoders and mobile telephones.

How can we effectively manage the complexities of these services with separate regulatory authorities each working within defined boundaries? The logic of the growing convergence of services and technologies is surely to combine the different statutes and regulatory bodies into common legislation under a single regulatory authority. In the early 1990s, media and telecommunications legislation was consolidated into three major Acts of Parliament governing telecommunications, broadcasting and radio communications, with an implicit agenda that these might eventually converge into a single *Communications Act*. The three major Acts were the *Telecommunications Act 1991*, which formalised the regulation associated with the Telstra–Optus duopoly and convergent forms such as pay television; the *Broadcasting Services Act 1992*, which changed the role of the Australian Broadcasting Authority away from a broadcasting policeman in some respects, fostered self-regulation and led to the ABA embarking upon pioneering governance work in another convergent form, the Internet; and the *Radiocommunications Act 1997*, which dealt with tradable convergent spectrum licences.

The possible shift towards a single industry regulatory authority for convergent communications (as has been adopted in Malaysia) was, however, undermined by the wider agenda of the perceived need for the Commonwealth government to implement national generic competition policy in 1995. The state governments accepted the argument for the different state utilities to be brought together under a common federal competition law, and the federal government agreed to subject telecommunications to the same law. Australia in the 1990s has essentially constructed regulatory structures and forms that have been driven by purely economic considerations. The great belief in the overall value of promoting competition as a vehicle for economic growth meant that the *Trade Practices Act 1974* assumed a new-found significance, and that the Australian Competition and Consumer Commission became the supreme government regulatory authority for all Australian industry—a body which initially had limited knowledge and expertise in media and communications and now grapples to keep up with its ever-expanding and tedious disputes workload. So, the institutional regulatory shift has been away from the prospect of the commonality of an industry-specific regulator for an age of convergence, towards the creation of a generalist regulatory body monitoring and promoting competition across every Australian industry. In the light of this key trend it is most unlikely that there will be convergence in communications legislation, with an associated single industry-specific regulatory body in Australia. The new power base of regulation is the ACCC, with its narrowly defined economic charter centred on competition policy, rather than the creation of a specialist body with a charter for monitoring and promoting the new dynamics and opportunities associated with convergence.

7. *We will need to maintain key regulation that enables us to prevent any further concentration of ownership in Australia's media.*

Convergence has not yet created a dynamic marketplace with many owners of different kinds. We cannot ignore the serious problem of concentration of ownership that still exists in Australian mainstream media. We do not yet have a plethora of players delivering a multiplicity of services that would enable us to abandon some key established media regulatory practices that have served useful national purposes.

The cross-ownership legislation that prevents the owner of a major Australian newspaper from also controlling the licence for a television channel ought to be maintained. During the inquiry into broadcasting conducted by the Productivity Commission in 1999, each of the major media corporations called for the abolition of the cross-media ownership restrictions. Commercial television interests have long defined competition policy in a self-interested way—they opposed the introduction of pay television for decades, and later suggested that there should be no more commercial television licences during the digital television policy review. Yet, they now see cross-ownership as an anachronism in competition terms. Murdoch's News Corporation now wants to keep its options open for a possible return to television in Australia but cannot do so while cross-ownership provisions apply to its 60 per cent of Australian newspaper ownership. Packer's PBL has long wanted to buy Fairfax for the lucrative classified advertising base of the *Age* and the *Sydney Morning Herald*, but they cannot do so while it remains in control of the Nine Network. Fairfax, with competition guru Fred Hilmer at its helm, could hardly show gross hypocrisy by calling for the retention of cross-ownership provisions, even though this legislation has almost certainly saved it from a Packer takeover in the past.

A credible case for changing the rules on cross-media ownership has not been made. The communication revolution is making its mark, but not to the extent that we now have demassification of the media with a plethora of new players. We have seen that commercial television and mainstream newspapers still command the heartland of Australian media audiences. The Internet is growing, but it is a beast of a different kind and its access remains restricted for many people. The removal of cross-media rules could lead to two companies, Packer's PBL and Murdoch's News Corporation, dominating 70 per cent of Australia's media marketplace. It is, of course, theoretically possible that a new entrant could buy Fairfax if the rules changed, but this is highly unlikely. The great danger is that the abolition of the cross-ownership provisions in broadcasting would unleash a comparable sequence of totally unproductive takeovers in commercial television similar to those in the late 1980s. The national interest must remain a higher priority than particular corporate plans for growth.

Long term, it is desirable that those responsible for framing our new media policies attempt to try to get not only more players into the marketplace but also greater ideological diversity into the system. Surely this is the promise that the new media systems hold for us. The promise of the new media is that a broader cross-section of our society will be involved in the managerial process of content construction, that there will be wider access to media outlets, and that there will be more media outlets of different ideological kinds.

8. *We must use the many vehicles of the new media for cultural development and promote content innovation in all of its different forms.*

Discussions about the development of 'electronic superhighways' tend to centre on high-speed, high-capacity information networks. The unease in the analogy between these communications highways and road traffic highways is that we were confident that the increased volume of cars would eventually come to justify the new highway construction. Yet, with the communications superhighways there is a notable lack of ideas about where all of the new content is going to come from to travel on these expensive new highways to support their existence.

We need to relate our ideas for the development of content to the outcomes of competition policy. Australian public policy holds different positions on competition policy for media and telecommunications. While telecommunications has become highly competitive in recent years, with a plethora of new carriers, commercial over-the-air television policy has remained under the control of the three networks. This policy has brought with it some content and cultural benefits which might otherwise not have been realised. The networks were guaranteed their good profits in a restricted three-player commercial television marketplace and the regulatory system could trade-off some indigenous content benefits.

Sections of the *Broadcast and Television Act 1942* held broadcasters responsible for providing 'adequate and comprehensive' programs, although regulators struggled to interpret what this meant or how it might be implemented. Regulators have also attempted to encourage specialised programming for a range of different audiences, such as children, or ethnic or linguistic minorities, although attempts to provide for program specialisation through regulation have generally not met with willing support from the commercial media. The introduction of a special category of television programs for children in the late 1970s, known as the 'C' category, whereby a specialist children's television committee recommended suitable children's programs for screening between 4 and 5 pm on weekdays, was an attempt to meet the needs of a special group of viewers. Although the networks disliked this form of program coercion, it contributed towards a diversity in children's programming that would otherwise not have existed.

Attempts have been made to foster national production, both in film and broadcasting, although this policy goal has been less consistent in recent years. The economics of purchasing commercial television programs favour imported programs over those locally produced by a factor of at least six to one. So, if *only* financial criteria had been applied by the Australian television networks in purchasing programs, it is unlikely that any Australian-made television programs would have been produced. Various regulatory mechanisms have been used to coerce the networks into screening local productions, including a points system for Australian content or a quota system for new Australian drama. These modes of regulation were devised to ensure that Australian production had some measure of access to television production outlets. Although widely opposed by network television management, it is unlikely that much Australian television drama would have appeared on Australian screens, and subsequently been exported, without such productive regulatory policies.

The present Australian Content Standards set the overall Australian programming on commercial television at 55 per cent between 6 am and midnight. The quota of C-classified children's programming is 260 hours a year, of which 130 hours must be first release Australian material. These regulations have served as a significant fillip to promote indigenous content through Australian television production companies, although new economic and trading rules undermine this commendable objective. For instance, in 1998 the Australian High Court ruled that these content standards contravened Australia's obligations under the Australia/New Zealand Closer Economic Relations Treaty (CER) trade agreement in the *Project Blue Sky* case. The court ruled that the regulator, the Australian Broadcasting Authority, was obliged to carry out its functions having regard to any international agreement to which Australia is a party. The ABA then amended its standards to accord national treatment to New Zealand programs. Local industry concerns are that this will detrimentally impact on Australian television production because Australian networks will be able to buy more New Zealand programs at secondary market prices to reach their redefined 'local' quota.

Those who believe that an unfettered commercial television marketplace will produce what audiences want, ought to look at the disastrous programming consequences that followed the deregulation of New Zealand commercial television. New Zealanders love to watch their talented rugby players on live-to-air television. The 1995 World Cup final, and the second Bledisloe Cup test against Australia, were watched live to air by over a million viewers in New Zealand. But when the brilliant All Blacks won their gladiatorial contest against South Africa for the first time in 1996, the nation could not indulge in their favourite nationwide television bonding ritual. By then TVNZ had lost the rights to live coverage, and this unforgettable contest was only broadcast live to air by the Sky TV pay television network. TVNZ could not show the match until twelve hours after this classic All Blacks–Springbok test.[12] Today, the New Zealand live-to-air program diet is dominated by cheap imports; local news and current affairs has lost its sense of substance; and commercials can now run for twenty minutes per hour. There are no longer any major sporting events broadcast live to air. Australia's anti-siphoning legislation for major sporting events prevents this kind of occurrence, but these rules are under attack from pay television interests and are likely to be progressively watered down. Is the New Zealand model of deregulation of broadcasting what Australia wishes to emulate in the future?

In the 1990s the political climate has shifted away from content/cultural heritage forms of regulation in view of a new-found trust in the benefits of market liberalisation which is supposed to 'let a thousand flowers bloom'. The new emphasis stresses that the fundamental task of regulators is to ensure the unfettered commercial operation of all players in the marketplace. The Australian Broadcasting Authority has had to move away from forms of traditional broadcasting regulation into ensuring that 'the marketplace' can achieve its objectives. Hence, the ABA has tended to become an overseer of structural regulation—similar to its more senior companion regulator, the Australian Competition and Consumer Commission—determining matters such as whether particular interests have an undue control in a broadcasting company due to their particular level of ownership. We have seen this with Can West in the case of Channel 10 as outlined in Chapter 3, and in issues about whether Packer exercised undue control through board representation in the Fairfax organisation. Is this what we want a broadcasting regulatory body to do? Perhaps we ought to talk about e-regulation rather than deregulation—'e' for *effective* regulation.

Do we intend to maintain national cultural objectives in broadcasting? Surely we ought to foster the development of indigenous television and film production in a world so hungry for product. Catherine Robinson has pointed out that the Australian film and television industry, by any measure, has enjoyed considerable success in recent years. Locally produced television programs enjoy consistently high ratings, and export revenue from the licensing of Australian television programs increased by 133 per cent between 1995 and 1998. 'Neighbours' has screened on the BBC for the past decade; and 'Neighbours' and 'Home and Away' between them have sold to 120 countries, 'Water Rats' to 169, 'Murder Call' to 100 and 'Blue Heelers' to 76. Australia now exports its children's television programs to 100 countries, and 'Bananas in Pyjamas', 'Johnson & Friends' and 'Blinky Bill' are much-loved characters throughout the world. In film we have a good international reputation; 144 Australian feature films were screened in 1997 at 72 international festivals and won 56 awards.[13]

The new media marketplace is one where convergence had fostered greater vertical integration of companies and a concentration of control around seven global entertainment corporations. Australian film production has enormous potential in an increasingly globalised film and television marketplace.

Yet the Australian film industry has long been handicapped by inadequate finance to support local productions, most of which have been low-budget films. Why does our brilliant film industry continually have to fight for survival? We have great talent in Australian film and television production, but no strategic national local production strategy to build on this performance. Governments have been more concerned with trying to devise taxation rules for film investment that would not embarrass them politically, than with attempting to secure the long-term business and cultural potential of an indigenous production industry. We seem to find it easier to invest long term in communications infrastructure than in communications content. Australia *can* afford a generous level of taxation incentives to support local film and television production to stabilise this industry. We must think more strategically about supporting the great creative attributes of Australians.

9. *A principal way in which government can act as a catalyst to promote local communications is by being an IT purchaser and user itself.*

Over many years, Australian governments have invested in various forms of assistance to promote local IT development in the form of R&D grants, loans and tax breaks. Yet no government, conservative or Labor, has had a coherent local purchasing policy to buy from the same companies in which they invested development capital. As a consequence, Australia has a serious $12 billion IT annual deficit. Moreover, the federal government, traditionally the largest purchaser of IT in the country, has moved to outsourcing its IT under the Howard administration. Essentially, this means that the government has abrogated its responsibility for IT purchasing to someone else who it pays for the service delivery. John Ride, CEO of Jtec, a medium-sized Australian telecommunications supply company, points out that 'indigenous IT companies are effectively shut out of the outsourcing tender process in favour of the transnationals . . . outsourcing leads to an 80% reduction in revenue for indigenous IT companies'.[14] What is the sense of maintaining this outsourcing policy?

More positively, to promote the on-line economy, one state government offers its citizens access to information about government and public utility services. The Victorian government has developed a system to put government services on-line in an Electronic Services Delivery (ESD) project using the Maxi product, widely agreed to be a world first. Maxi Multimedia Pty Ltd is an ESD joint venture between NEC Australia and Aspect Computing, which provides people with electronic access to various government and business agencies for information and services. Maxi delivers its services 24 hours a day, seven days a week, through different channels via a touch tone telephone, the Internet or through multimedia kiosks at popular locations. It is possible to make payments for rates, licence fees, fines, and so on, as well as booking venues, seats at events and licence tests, if you wish. This offers many innovative and useful services, such as people being able to advise a multiplicity of agencies of their change of address just once through an on-line channel, instead of having to separately and painstakingly contact gas, water, electricity, post and telephone companies, as well as the electoral commission. By June 1999 Maxi was processing 40 000 transactions per month, 40 per cent of which were outside of normal business hours. Maxi claims to be both safe and secure through authentication software, and only the agencies authorised by the user will have access to personal information—which will not be retained by Maxi. While citizens will no doubt eye Maxi with understandable vigilance, the project shows refreshing initiative on the part of a government leading by example in its commitment to building an on-line culture.

10. *We need invest in fundamental research in communications.*

The unprecedented levels of investment in information networks, products and services, and the vast range of new communications forms, offer us a cyber-sea of information. We are increasingly dependent on information to run our lives. Some sceptics argue that many aspects of the new communications systems are actually a barrier to effective communication. In fact, we know little about the effects of new communication practices and processes on our lives.

It is extraordinary how little fundamental research is undertaken within the Australian communications industry. Given the explosion of new services that are now being pumped into an overcrowded media marketplace, it is astonishing that so many companies would chance their arm in the hope that the product or service will eventually get some kind of marketplace take-up. At the end of the day, the only communication products and services that will work commercially are those that fill needs and are affordable. The success of the range of new on-line services will primarily depend upon how the organisations that offer those services understand the behavioural practices and needs of consumers and citizens. The real question for our communication future is not what the *technologies* are going to be like, but what *we* are going to be like.

We do have plenty of marketing-based research about consumer choices, product preferences and perceived quality of service—all very necessary in our new age of cut-throat competition. But there is no fundamental research being undertaken in Australia in communications. There are plenty of questions we urgently need to explore. It is remarkable that we have embraced the paradigm of competition policy, and yet we have initiated almost no research into its outcomes. Where is the evidence based on research that the full privatisation of Telstra would lead to a more efficient and responsive national communications carrier? What do we really know, too, about the possible emergence and impact of an 'on-line society'? How do we know whether the demand for these new information services is sufficient to justify the substantial levels of development investment? Will it be more efficient, and better for the community, if we experience a major shift in informational transactional practices by moving towards widespread acceptance of on-line banking, virtual education, electronic commerce, telemedicine, home shopping, home gambling and video on demand? What are the social implications for Australia that might follow widespread deinstitutionalisation of banking, retailing, health, education and entertainment?

We need more than applied or product-based research. Surely this $70 billion industry can afford to invest in *fundamental* research. Many digital technologies have infiltrated our lives, and appear likely to have an even greater effect in the future. Little systematic attention has been given to how the changes of the information society impact on our lives—the biggest questions are rarely asked, let alone explored. How do all of these changes affect the way we communicate with each other, individually and collectively? Is our traditional print-based culture now becoming an electronic culture? Will we be better able to communicate with each other? Are our lives richer as a result of these changes?

These are some of the fundamental research agendas for Australian communications in the future.

Whither Australia's communications?

So, whither Australia's communications policy? Some of the major debates of the new century will centre on resolving the tensions between the shift towards a highly international privatised economy, with more liberalised domestic communications industries, but where the issues of access, equity, education and social justice must be addressed. Australia will need to ensure that it has a communications industry that is internationally competitive and of world's best practice. We need private investment in productive enterprises that successfully trade globally out of Australia. The pendulum is likely to shift back towards closer examination of areas designated as 'market failures' and towards a re-evaluation of the significant role that better integration of public and private sector practices can have. We need to contribute more to public debates about the so-called third way—the renewal of social democracy—and articulate the renewed legitimacy for the public sector. History may well repeat itself in terms of the key role of the public sector. The state still has critical responsibilities in establishing regulatory roles that foster a fair but competitive communications environment, in planning long-term infrastructure projects and policies of national significance to ensure equity, in fostering an innovative culture of research and development, in developing employment-generation programs, in supporting life-long learning, and in helping those in geographically isolated areas and with disabilities. We ought to find ways to provide support and opportunities for brilliantly creative Australians who do not fit within the intensely market-driven new media paradigm.

So, what is our ultimate future perspective for the communications society? Australia's communications environment has changed more during the past five years than during any other time in its history, and there are few signs that the scale and rapidity of change is likely to abate. The implications for the new communications revolution are profound—they require astute national economic strategies, global positioning, the initiating of fundamental research, ensuring widespread community access, and exploitation of our comparative trading advantages in an intensely competitive world.

New media offer new ways of purchasing goods and services and of doing business, as well as changes in the ways we are educated, how we entertain ourselves, conduct research and generally organise our lives. Communications, in all of its new guises and forms, is contributing to unprecedented social change which might collectively represent an age of electronic or digital culture. Hopefully, the outcome will be that our lives are not so much digitally determined, but enhanced and enriched in this brave new world of communications.

It still remains up to us.

Endnotes

Preface

1 World Telecommunication Development Report, *Universal Access*, International Telecommunication Union, Geneva, 1998.
2 *Digital Planet: The Global Information Economy*, International Data Corporation, 1998, p. 5.
3 'Internet drives Telstra's future network spend', *Telstra Press Release*, 8 December 1998.
4 'Knowledge for Development', *World Development Report*, World Bank, Washington, DC, 1998/99, p. 24.

1 Media moguls: Power, ownership and influence

1 Henry Mayer, *The Press in Australia*, Lansdowne Press, Melbourne, 1964; Keith Windchuttle, *The Media*, Penguin Books, Melbourne, 1998; and Annual Reports.
2 Report of the *Inquiry into the Ownership and Control of Newspapers in Victoria*, by Hon. J.G. Norris, Report to the Premier of Victoria, 15 September 1981.
3 Vic Carroll, 'Maintaining a cranky media', *Australian Financial Review*, 11 March 1997, p. 18.
4 Peter Westerway, Conference paper, Communications Law Centre, Sydney, April 1997, p. 41.
5 Jock Given, CLC Conference, op. cit., p. 72.
6 R. Walker, *Yesterday's News: A History of the NSW Press from 1920 to 1945*, Sydney University Press, Sydney, 1980, quoted by Julianne Schultz in *The Media in Australia*, eds S. Cunningham and G. Turner, Allen & Unwin, Sydney, 1997, p. 32.
7 Michael Duffy, CLC Conference, op. cit., p. 9.
8 Rod Tiffen, 'Packer: The timing is nigh', *Australian Financial Review*, 9 May 1997.
9 C. Boggs, *Gramsci's Marxism*, Pluto Press, London, 1976, p. 39.
10 J. Lull, *Media Communication and Culture*, Columbia University Press, New York, 1995.
11 Roy Morgan research, *Australian Financial Review*, 3 August 1999.
12 Based on A.C. Nielsen's 1997 Trend Statistics, and Nielsen data published in the *Australian Financial Review*, 1 August 1999.

13 Margaret Easterbrook, 'Net surf's up at TV's expense' (data of the Newspaper Advertising Bureau of Australia), *Age*, 14 May 1998, p. A3.
14 Paul Chadwick, 'Read all about it less', *Communications Update*, April 1997, p. 12.
15 Easterbrook, op. cit.

2 Forces for change: Communications as catalyst

1 'The cutting edge', *Australian*, 6 April 1999, p. 3.
2 Greg Callaghan, 'Total telephone mobility', *Australian*, 29–30 May 1999.
3 Frank Blount, 'Telecoms: Coming to grips with open markets', *Asian Media and Communications Bulletin*, vol. 27, no. 4, p. 10.
4 R. Winsbury, 'How grand are the alliances?', *Intermedia*, vol. 25, no. 3, 1997, p. 27.
5 J. Tomlinson, 'Mass communication and the idea of a global public sphere', *Journal of International Communication*, July 1994.
6 Kelvin Rowley, 'Boom, bust and beyond: The Thai crisis in the global perspective', paper, Mekong Perspectives Conference, Australian National University, 22 March 1999.
7 Roy Allen, *Financial Crises and Recession in the Global Economy*, Edward Elgar, London, 1994, p. 1.
8 US Government, *Economic Report of the President, Transmitted to Congress*, US Government Printing Office, Washington, DC, 1999, p. 224.
9 Manuel Castells, *The Information Age: Economy, Society and Culture* (series), vol. 1, *The Rise of the Network Society*, Blackwells, Oxford, 1996, p. 472.
10 ibid.
11 Les Macdonald, *Australian Rationalist*, no. 47, 1998, p. 51.
12 'Avoid global trap', *Australian Financial Review*, 23 March 1998, p. 16.
13 A. Giddens, *A Contemporary Critique of Historical Materialism: The Nation States and Violence*, vol. 2, University of California Press, Berkeley, 1995, p. 120.
14 Castells, 1996, op. cit., p. 258.
15 Manuel Castells, *The Information Age: Economy, Society and Culture* (series), vol. 2, *The Power of Identity*, Blackwells, Oxford, 1997, p. 243.
16 ibid., pp. 258–9.
17 Anthony Smith, *Goodbye Guttenberg* (video), 1980.
18 Castells, 1996, op. cit., p. 18.
19 Alain Benoist, 'Confronting globalisation', *Telos*, no. 108, Summer 1996, p. 121.
20 ibid., pp. 121–2.
21 ibid., p. 123.
22 Brian Toohey, *Tumbling Dice*, William Heinemann Australia, Melbourne, 1994, p. 217.
23 'Will telecommunications remain a golden sector?', *Telecommunications Policy*, December 1992.

3 Challenge of change: Australia's media institutions

1 Quoted in Alan Deans, 'Fox trading 20pc below par', *Australian Financial Review*, 14 December 1998.

2. Alan Kohler, 'Go figure: News is flying', *Australian Financial Review*, 20–21 June 1998.
3. Cited in Ten National Holdings Ltd Prospectus, 4 March 1998, based on data provided by FACTS.
4. John Button, *As It Happened*, The Text Publishing Company, Melbourne, 1998, p. 227.
5. Peter Westerway, 'Twenty years ago', paper presented to the Media Ownership and Control Conference, Communications Law Centre, Sydney, 10 April 1997.
6. Jane Schultze, 'PBL Online float plan', *Age*, 3 February 1999.
7. Mark Day, 'Late news on the Internet', *Australian*, 1–7 April 1999, Media supplement, p. 8.
8. C. Mathieson and J. Dunbar, 'We're in Fairfax for the long haul: BIL', *Australian*, 26 July 1998.
9. Quoted in Peter Witts, 'Cost cutting saving Fairfax millions', *Weekend Australian*, 7–8 November 1998.
10. Kevin Morrison, 'Fairfax to focus on building online muscle', *Age*, 8 February 1998, Business, p. 1.
11. Alan Kohler, 'Net trawls the rivers of gold', *Australian Financial Review*, 1–5 April 1999, p. 29.
12. Luke Collins, 'Seven: More radical changes', *Australian Financial Review*, 19 March 1999, p. 53.
13. G. Butlar, 'Intel chips in with Seven online', *Australian Financial Review*, 25 March 1998.
14. Jane Schultze, 'Seven's online onslaught', *Age*, 28 July 1999, Business, p. 6.
15. R. Myer, 'Channel's amazing tale of revival', *Sunday Age*, 29 March 1998.
16. ibid.
17. S. Anderson, 'Australis adds $297m to pay-TV losses', *Australian Financial Review*, 12 September 1997.
18. Finola Burke, 'Australis Media in the hands of receivers', *Australian Financial Review*, 6 May 1998.
19. S. Lewis and F. Burke, 'The $3 billion pay-TV fiasco', *Australian Financial Review*, 29 August 1997, p. 1.
20. Company data. Australia has seven million households, according to the Census of Population and Housing, Australian Bureau of Statistics, 1996.
21. Minister for Finance, Telstra Share Offer, Public Offer Document, September 1997, p. 45.
22. *1998 Australian Domestic Media Sector Report*, J.B. Were & Son, February 1998, p. 15.
23. Finola Burke, 'Packer plays end game', *Australian*, 27 March 1999.
24. Finola Burke, 'PBL joins News in Foxtel', *Australian Financial Review*, 4 December 1998.
25. Dennis McQuail, *Media Performance: Mass Communication and the Public Interest*, Sage, London, 1992, p. 2.
26. K.S. Inglis, *This is the ABC*, Melbourne University Press, Melbourne, 1983, p. 5.
27. ibid., p. 25.
28. Ien Ang, *Desperately Seeking Audiences*, Routledge, London, 1991, pp. 109–10.
29. ibid.
30. L. Goode, 'Media systems, public life and the democratic project', *Arena*, no. 7, 1996, p. 68.

31 'Survey shows audiences place a very high value on the ABC', *ABC Media Release*, 11 February 1999.
32 ABC Annual Report, 1996/97.
33 Marc Raboy, 'Public service broadcasting in the context of globalisation', in *Public Service Broadcasting for the 21st Century*, ed. M. Raboy, University of Luton Press, Luton, UK, p. 2.
34 Broadcasting Research Unit, *The Public Service Idea in British Broadcasting: Main Principles*, BRU, London, 1985.
35 B. Johns, excerpts of the submission to the *Review of the Role and Function of the ABC* (known as the Mansfield Review).
36 *The ABC in Review*, Report by the Committee of Review of the Australian Broadcasting Commission, AGPS, Canberra, 1981 (known as the Dix Report).
37 Robert Mansfield, *The Challenge of a Better ABC*, AGPS, Canberra, 1997, vol. 1, p. 39.
38 ibid. Recommendation no. 15, p. 37.
39 Quoted in G. Davis, 'The Mansfield Review: Cautious, conservative and supportive', *Communications Update*, Issue 130, March 1997.
40 *Age*, 21 September 1998.
41 K. Lenthall, 'Push for probe on ABC expands', *Age*, 21 September 1994.
42 Sir Rupert Hamer and June Factor, *Save our ABC, The Case for Maintaining our Independent National Broadcaster*, Hyland House, in association with Friends of the ABC, 1998.
43 Dix Inquiry, op. cit., Part 1.
44 Mansfield Inquiry, op. cit., p. 4.
45 SBS Annual Report, 1996/97, pp. 52 and 62.
46 'An Introduction to SBS Australia', at *http://www.sbs.com.au/faq.html#audiences*
47 R. Oliver, 'Can this man save SBS?', *Age Green Guide*, 2 April 1998.

4 Citizens to customers: Telecommunications in transition

1 See Ann Moyal, *Clear Across Australia*, Nelson, Melbourne, 1986.
2 ibid.
3 Mel Ward, 'Telecommunications and information industry and economic development', speech presented to the Science and Industry Forum, Sydney, 7 July 1987.
4 Quoted in Finola Burke, 'It's a hit: Optus floats into the top 10', *Australian Financial Review*, 18 November 1998.
5 Bureau of Industry Economics (BIE), *Information Technology and Telecommunications Industries: An Evaluation of the Partnerships for Development and Fixed Term Arrangements Program*, AGPS, Canberra, 1995, pp. 167–8.
6 Fred Hilmer et al., *National Competition Policy*, Report of the Independent Committee of Inquiry, AGPS, Canberra, 1993.
7 Jim Holmes, 'Aussie style interconnect—is it working?', paper presented to the Australian Telecommunications Users Group Conference, Melbourne, 6 May 1998.
8 ibid.

9. Grant Butler, 'Telstra ordered to cut call costs', *Australian Financial Review*, 20 January 1998, p. 1.
10. *Report of the Committee of Inquiry into Telecommunications Services in Australia*, vol. 1, AGPS, Canberra, 1982, p. 137.
11. Patrick Xavier, 'Australia's post July 1997 telecommunications regulation in the international context', paper presented to the BTCE Communications Forum, Melbourne, 28 October 1996.
12. ibid., p. 11.
13. Productivity Commission, 'Telecommunications economics and policy issues', *Staff Information Paper*, March 1997.
14. Allan Brown, 'Reform and regulation of Australian telecommunications', Working Paper no. 5, Griffith University, 1996, p. 22.
15. 'Evaluation of the transitional period in Australian telecommunications', BTCE Working Paper, 1995.
16. Brown, op. cit., p. 22.
17. 'Six month extension to Telstra price caps', *Media Release*, Department of Communications Information Technology and the Arts, 14 December 1998.
18. Warwick Smith, 'Telecommunications: The only constant is change', paper presented to the ATUG Conference, Melbourne, April 1998.
19. J. Muir, B. Jennings and F. McAnally, 'Price competition in Australian telecommunications', paper presented to the BTCE Forum, Canberra, 24–25 September 1998.
20. John de Ridder, 'International comparisons of telecommunications prices', paper presented to the BTCE Forum, Canberra, 24–25 September 1998.
21. *Review of the Standard Telephone Services*, DOCA, Canberra, 1996, p. 29.
22. Holly Raiche, 'Universal service in Australia', in *All Connected: Universal Services in Telecommunications*, ed. B. Langtry, Melbourne University Press, Melbourne, 1998.
23. Quoted in Brown, op. cit., p. 11.
24. *Reforming Universal Service: The Future of Consumer Access and Equity in Australian Telecommunications*, Consumer Telecommunications Network, Sydney, 1993.
25. An idea first proposed by David Sless at the Commission for the Future in 1987.
26. Quoted by Paul Kelly in 'Social shackles thwart Telstra', *Australian*, 23 December 1998, p. 13.
27. Telstra, *Annual Review*, 1998, p. 3.
28. Taken from C. Johnson, *Privatisation of Utilities: How are Consumers Affected?*, Australian Consumers Council Issue, AGPS, Canberra, 1995.
29. See *World Development Report*, ITU, Geneva, 1994, p. 56.
30. Steve Lewis, 'Telstra sale on the fast track', *Australian Financial Review*, 21 June 1999, p. 1.
31. See 'Telstra Privatisation', Background Briefing Paper, Communications Electrical and Plumbing Union (CEPU), June 1998.
32. Allan Brown, 'Should Telstra be privatised?', paper presented to the 24th Conference of Economists, Adelaide, 24–27 September 1995, p. 2.
33. *Telstra—To Sell or Not to Sell*, Senate References Committee, AGPS, Canberra, September 1996.
34. S. Domberger, 'The role of public enterprise in microeconomic reform', in *Microeconomic Reform in Australia*, ed. P. Forsyth, Allen & Unwin, Sydney, 1992, pp. 168–9.

35 Dianne Northfield, 'Weighing up the arguments in the Telstra privatisation debate', unpublished paper, 25 November 1995.
36 OECD data quoted by Henry Ergas in 'Four Corners', 5 July 1989, and see also *International Performance Indicators: Telecommunications 1995*, BTCE, AGPS, Canberra, 1995.
37 S. Domberger and J. Piggot, 'Privatisation policies and public enterprises: A survey', *Journal of the Royal Statistical Society*, vol. 148, part 2, 1995, p. 145.
38 J. Vickers and G. Yarrow, 'Economic perspectives on privatisation', Journal of Economic Perspectives, Spring 1991, p. 12.
39 'Telstra poll push', *Age*, 16 March 1998, p. 1.
40 'Telstra sale to cut debt to $25 billion', *Australian Financial Review*, 24 February 1998, p. 7.
41 Frank Blount, 'Telstra calling out for full privatisation', *Australian Financial Review*, 1 March 1999.
42 John Quiggin, 'Telstra float sounds too good', *Australian Financial Review*, 19 March 1998.
43 Kenneth Davidson, 'No sense in the Telstra sell-off', *Australian Financial Review*, 21 June 1999.
44 Steve Lewis and Louise Dodson, 'Telcos connect to fight further Telstra sale', *Australian Financial Review*, 22 March 1999.
45 See views expressed at the 1998 ATUG Conference, Melbourne, May 1998, as reported in *Australian Financial Review*, 6 May 1998, p. 27.
46 Steve Lewis, 'Connecting to Telstra costs too much', *Australian Financial Review*, 7 December 1998.
47 Helen Meredith, '"Remember the customer", says ATUG', *Australian Financial Review*, 31 March 1998, p. 27.

5 Electronic nomads: Internet as paradigm

1 Thomas S. Kuhn, *The Structure of Scientific Revolutions*, 2nd edn, University of Chicago Press, Chicago, 1970, p. viii.
2 See American Civil Liberties Union at http://www.aclu.org/coutt/cdadec.html
3 John Hindle, Vice President, Nortel, in *The Internet as Paradigm*, Institute for Information Studies, Aspen Institute, 1997.
4 Chief Judge Slovitor in ruling the US *Communications Decency Act* unconstitutional, 11 June 1996.
5 Schweitzer, quoted in Manuel Castells, *The Rise of the Network Society*, Blackwells, Oxford, 1996, p. 359.
6 ibid., p. 360.
7 See 'Studies reveal typical Net user', Office of Women's Affairs, at *infolink@thehub.com.au*
8 Australian Bureau of Statistics, 'Home Internet use grows strongly', *Media Release*, 130/98, 25 November 1998.
9 Australian Bureau of Statistics, 'Household Use of Information Technology', ABS Cat. No. 8146.0, May 1999.
10 American Civil Liberties Union; see http://www.aclu.org/court/cdadec.html
11 Robert Hobbes Zakon, *Hobbes' Internet Timeline*, The Mitre Corporation, 31 December 1998, p. 14.

12 Reported by Ellen Fanning, *The World Today*, 21 April 1998.
13 Mitchell Kapor, 'Where is the digital highway really heading?', *Wired*, July/August 1993.
14 ibid.
15 John Hindle, Vice President, Nortel, in *The Internet as Paradigm*, Institute for Information Studies, Aspen Institute, 1997.
16 Nicholas Negroponte, *Being Digital*, Hodder & Stoughton, Sydney, 1995, p. 229.
17 J.P. Barlow, quoted in *The Governance of Cyberspace*, ed. B.D. Loader, Routledge, London, 1997, p. 4.
18 Jon Katz, *Wired*, June 1995.
19 Howard Rheingold, *The Virtual Community*, Addison-Wesley Publishing, London, p. 1.
20 ibid., p. 3.
21 Sherry Turkle, *Life on the Screen*, Simon & Schuster, New York, 1995, p. 12.
22 J. Fernback and B. Thompson, 'Virtual Communities: Abort, Retry, Failure?', 1995. See *http://www.wel.com/user/hlr/texts/VCcivil.html*, p. 2.
23 ibid., p. 3.
24 Quoted in Loader, op. cit., p. 38.
25 Mark Slouka, *War of the Worlds*, Abacus, London, 1995.
26 Clifford Stoll, *Silicon Snake Oil*, Pan Books, London, 1996, p. 148.
27 R. Kraut, M. Patterson, V. Lundmark, S. Kiesler, T. Mukopadhyay and W. Scherlis, 'Internet paradox: A social technology that reduces social involvement and psychological well-being?, *American Psychologist*, September 1998.
28 ibid., p. 1028.
29 Quoted in Slouka, op. cit., p. 60.
30 Anthony Smith, *Goodbye Guttenberg*, BBC video, 1980.
31 Michael Cosgrove, ACCC, speaking at Internet Industry Association Conference, Melbourne, 8 December 1997.
32 Paul Robinson, 'Internet use by abusers rising, say investigators', *Age*, 14 September 1997.
33 K. Lawrence, 'Internet paedophile warning', *Sunday Herald Sun*, 26 October 1997.
34 See *http://www.anu.edu/people/Roger. Clarke/II/CensCope.html*
35 US District Court, *Eastern District of Pennsylvania, American Civil Liberties Union et al. v Janet Reno, US Attorney General*, Civil Action No. 96–963. See *http://www.aclu.org/court/cdadec.html*
36 ibid., p. 64.
37 See 'Development of professional on line services in China: Its obstacles and outlook', Proceedings of the Eighth Australasian Information On-line and On-disc Conference and Exhibition, Sydney Convention and Exhibition Centre, Sydney, 21–23 January 1997.
38 *The Internet and Some International Regulatory Issues Relating to Content*, a pilot comparative study prepared for UNESCO, Australian Broadcasting Authority, Sydney, 1997, p. 39.
39 ibid., p. 53.
40 R. O'Neale, 'Freedom of expression and community standards in the on-line world', Communications Research Forum, BTCE, Canberra, 1998.
41 Mark Forbes, 'www.sexlaws.not.on', *Age*, 8 June 1999, p. 15.
42 'National framework for the on-line content regulation', Senator Richard Alston,

Minister for Communications and Darryl Williams, Attorney General, *Joint Media Release*, 15 July 1997.
43 See *http://www.dca.gov.au/online/on-cont.html*
44 *http://www.whitehouse.gov/WH/NEW/Ratings/remarks.html* at 16 July 1997.
45 D. Crowe, 'Naughty Net providers put on criminal notice', *Australian Financial Review*, 13 December 1997, p. 1.
46 ibid.
47 This term is used by the Yellow Pages Small Business Index, Telstra, April 1998.
48 'Brands bite back', *The Economist*, 21 March 1998, p. 82.
49 See *E-Commerce: Our Future Today*, Andersen Consulting, April 1998, IDC's *Internet Commerce and Usage in Australia, 1997–2002*, cited at *http://www.nua.ie/surveys/*, and Department of Industry, Science and Tourism, *Stats: Electronic Commerce in Australia*, Commonwealth of Australia, Canberra, 1998.
50 IDCs quoted in the *Australian*, 23–24 May 1998, US Congress, 'Realising the Electronic Commerce Revolution', 1998.
51 Australian Department of Foreign Affairs and Trade, *Putting Australia on the New Silk Road: The Role of Trade Policy*, 1997, and IDC at *http://www.idc.com/*
52 Electronic Commerce Expert Group, *Electronic Commerce: Building the Legal Framework*, Report to the Attorney General, March 1998.
53 M. Krantz, 'Click till you drop', *Time*, 20 July 1998, p. 44.
54 ibid., p. 31.
55 ibid.
56 Hoovers OnLine, 'Hoovers Company Profiles: Amazon.com Inc', at *http://www.hoovers.com*, reviewed 28 July 1998.
57 'Net stocks: Short cut to what may be a short lived fortune', *Australian Financial Review*, republished from the *Wall Street Journal*, 6 January 1999.
58 Marc Phillips, *Successful E-Commerce*, Bookman, Melbourne, 1998.
59 See Table, *Australian Financial Review*, 23 December 1998, p. 40.
60 Bruce Jacques, 'Cyber crazy', *Bulletin*, 16 February 1998, p. 36.
61 *E-commerce: Our future today*, op. cit.
62 Peter Gerrand, 'Commerical Internet Domain Administration in Australia', *Telecommunications Journal of Australia*, vol. 48, no. 3, 1998.
63 Jon Katz, 'Birth of a digital nation', *Wired*, April 1997.

6 Being human: Paradoxes of the new media

1 Peter Large, *The Micro Revolution*, Fontana, London, 1980, p. 9.
2 Christopher Evans, *The Mighty Macro*, Coronet, London, 1979, p. 94.
3 Tom Stonier, *The Wealth of Information: A Profile of the Post-Industrial Economy*, Thames Methuen, London, 1983.
4 Hans-Peter Martin and Harald Schumann, *The Global Trap*, Zed Books, New York, 1997.
5 Marian Marien, 'New communications technology', in *Telecommunications Policy*, vol. 20, no. 5, 1996, p. 375; Z. Sarder and J. Ravetz (eds), 'Cyberspace: To Boldly Go', *Futures*, September 1996.
6 *World Telecommunication Development Report*, International Telecommunications Union, Geneva, 1998, p. 72.
7 Ann Moyal, *Clear Across Australia*, Nelson, Melbourne, 1988.

8 *Report of the Commission of Inquiry into the Australian Post Office*, vol. 1, April 1974, pp. 36–8.
9 *Review of the Standard Telephone Services*, DOCA, Canberra, 1996, p. 29.
10 J. St Clair, J. Muir and A. Walker, 'Digital technologies in Australian homes', paper presented to the Communications Research Forum, Melbourne, 28 October 1996.
11 Australian Bureau of Statistics, 'Home Internet use grows strongly', *Media Release*, 130/98, 25 November 1998.
12 'Gulf widens as the poor get poorer', *Age*, 24 March 1998, News, p. 6.
13 This concept emerged in the work of the BTCE, *Communication Futures*, AGPS, Canberra, 1995, p. 140.
14 Extracts from Telstra Annual Report, 1997, p. 5.
15 ibid., p. 10.
16 Sandy Kyrish, 'Public policy and the cycle of prediction', paper presented to the BTCE Conference, Melbourne, October 1996.
17 Howard Segal, *Technological Utopianism in American Culture*, University of Chicago Press, Chicago, 1985.
18 *Communications Futures*, BTCE, 1994, p. 70.
19 D. Brewster, 'Telstra wears all the pain ahead of November float', *Australian*, 30–31 August 1997, p. 53.
20 Telstra Annual Report, 1998, p. 10.
21 Frank Blount, Charles Todd Memorial Lecture, University of Melbourne, 2 September 1997.
22 *Australia's Networking Future*, BSEG, 1995, p. 76.
23 George Gilder, *Life After Television*, Norton, New York, 1994.
24 Paul McIntyre, 'Only some survive media turmoil', *Australian Financial Review*, 26 September 1995, p. 42.
25 'Technology and entertainment supplement: Wheel of fortune', *The Economist*, 21 November 1998, p. 3.
26 ibid., p. 10.
27 Neil Postman, *Technopoly*, Vintage Books, New York, 1993.
28 *International Business Week*, 29 April 1997, pp. 36–41.
29 David Sless, *Sydney Morning Herald*, supplement on Communications, June 1990.
30 Quoted in Maureen Mackenzie-Taylor, 'Designing for understanding within a context of rapidly changing information', paper presented to the Vision Plus 3 Conference, Vorarlberg, Austria, July 1997, p. 1.
31 Brenda Dervin, 'Users as research inventions: How research categories perpetuate inequalities', *Journal of Communication*, vol. 39, no. 3, 1989, p. 216.
32 Quoted in Sless, op. cit.
33 'Technology and entertainment supplement: Wheel of fortune', op. cit., p. 3.
34 Anura Goonasekera and Duncan Holaday (eds), *Asian Communication Handbook*, AMIC, 1998, p. x.
35 Anthony D. Smith, 'Is there a global culture?', *Intermedia*, vol. 20, no. 4–5, 1992.
36 Speech given by Rupert Murdoch to advertisers in London, 1 January 1993; extracts shown on 'Sunday', Nine Network.
37 Herbert Schiller, 'Transactional media and national development', in *Media Culture and Regulation*, ed. Kenneth Thompson, Sage, London, 1997, pp. 154–5.
38 Brian Shoesmith, 'Technology transfer or technology dialogue: Rethinking western communication values', in Goonasekera and Holaday, op. cit., p. 274.

39 See 'Cultural markets in the age of globalisation: An Asian perspective', *Inter Media Special Report*, December 1997.
40 See D. Gautier, AMIC Conference, Jakarta, 24 June 1995.
41 See 'Media spectatorship', in *Global Local*, eds R. Wilson and W. Dissanayake, Duke University Press, North Carolina, 1996, p. 148.
42 ibid.
43 E. Herman and R. McChesney, *The Global Media*, Cassell, London, 1997, p. 44.

7 Towards an information society

1 See J. Carey, *Communication as Culture: Essays on Media and Culture*, Unwin Hyman, London, 1989.
2 Ithiel de Sola Pool, *Technologies of Freedom*, Harvard University Press, Cambridge, MA, 1983.
3 Daniel Bell, 'The social framework of the information society', in *The Microelectronics Revolution*, ed. T. Forester, Basil Blackwell, London, 1980, p. 501.
4 See M. Connors, *The Race to the Intelligent State: Towards the Global Information Economy of 2005*, Blackwells, Oxford, 1993; P. Drucker, 'The age of social transformation', *The Atlantic Monthly*, 1994, p. 274; A. Toffler, *The Third Wave*, William Morrow & Co., New York, 1995; F. Koelsch, *The Infomedia Revolution, How it is Changing our World and your Life*, McGraw-Hill and Ryerson, Toronto, 1995; S.G. Jones (ed.), *Cybersociety: Computer Mediated Communication and Community*, Sage Publications, London, 1995; and M. Castells, *The Rise of the Network Society*, Blackwell Publishers, Oxford, 1996. Summarised in M. Marien, 'New communications technology', *Telecommunications Policy*, vol. 20, no. 5, 1996, p. 375.
5 Arthur Andersen, *1998 Report on the Communications, Media and Entertainment Industries*, quoted in *Australian Financial Review*, 7 August 1998.
6 *Digital Planet: The Global Information Economy*, International Data Corporation, 1998, p. 5.
7 *Information Technology Outlook 1997*, OECD, Paris, 1997, p. 14.
8 Al Gore speaking at the First World Telecommunications Development Conference, Buenos Aires, 21 March 1994.
9 ibid.
10 Quoted in Clifford Stoll, *Silicon Snake Oil*, Macmillan, London, 1995.
11 US Office of the Vice President, *The Global Information Infrastructure: Agenda for Cooperation*, US Government Publishing Office, Washington, DC, 1995.
12 'Europe and the global information society: Recommendations to the European Council', Brussels, 26 May 1994.
13 Outline of 'Vision 21 for info-communications: Policies and practicalities for the 21st century', *New Breeze*, vol. 9, no. 3, July 1997.
14 ibid., p. 3.
15 *A Vision of an Intelligent Island, IT 2000 Report*, March 1992, National Computer Board, Singapore, p. 19.
16 See *http://www.s.one.gov.sg/SING.ONE/NC3.23E.Html*
17 'An invitation to Malaysia's MSC: Leading Asia's information age', Kuala Lumpur, 1997.

18 The APEC Ministerial Meeting on Telecommunications and Information Industry, Seoul, 29–30 May 1995.
19 *Australian Information and Communication Services and Technologies: Compendium of Legislative Reviews and Related Research Activities*, CIRCIT, Melbourne, 1995.
20 As quoted in 'Drivers of development in information and communication services: International experience and directions', Draft Paper, CIRCIT, Melbourne, 1997, p. 2.
21 ibid., p. 3.
22 T. Mandeville, G. Hearn, D. Rooney, J. Foster, T. Stevenson and D. Anthony, 'Knowledge model', paper presented to the BTCE Conference, Canberra, October 1998.
23 'The limits to IT', *Australian Financial Review*, 29 July 1997.
24 World Bank, *World Development Report: Knowledge for Development*, 1988/89, World Bank, Washington, DC, p. 19.
25 Information Industries and Online Task Force, *Stocktake of Australia's Information Industries*, Report prepared by STM Consulting Pty Ltd, Department of Industry, Science and Tourism, p. vii.
26 ibid., p. 53.
27 *The Global Information Economy: The Way Ahead*, Information Industries Task Force, July 1997, p. 48.
28 *Spectator or Serious Player? Competitiveness of Australia's Information Industries*, The Allen Consulting Group Pty Ltd, March 1997.
29 *The Global Information Economy*, op. cit., p. 104.
30 *Spectator or Serious Player?*, op. cit., Chapter 1, p. 23.
31 'Big Blue and Big Brother', *Australian Financial Review*, 5 October 1997, p. 31.
32 *Going for Growth: Business Programs for Investment, Innovation, and Export*, DIST, June 1997, p. 12.
33 *The Global Information Economy*, op. cit., p. 84.
34 Quoted in 'Submission to the review of business programs', AEEMA, Canberra, March 1997, p. 10.
35 ibid.
36 Peter Roberts, 'Industry sings electronic blues', *Australian Financial Review*, 12 August 1998, p. 14.
37 ibid.
38 Clive Mathieson, 'Internet upsurge spurs Telstra', *Age*, 24–25 October 1998.
39 C. Eugster, G. Besio and J. Hawn, 'Builders for a new age', *The McKinsey Quarterly*, no. 3, 1998, New York, p. 92.
40 *Spectator or Serious Player?*, op. cit.
41 Tim Colebatch, 'New call for technology investments', *Age*, 25 August 1997, p. 1.
42 *The Global Information Economy*, op. cit., p. 104.
43 D. Crowe, 'Intel deal not cheap as chips', *Australian Financial Review*, 4–5 May 1998, p. 1.
44 *Make or Break: 7 Steps to Make Australia Rich Again*, MTIA/EIU, 1997, p. 15.
45 M. Grattan, 'Whose baby is IT anyway', *Australian Financial Review*, 20–21 September 1997, p. 26.

8 Whose vision?: Third way communications

1 Robert Reich, *The Work of Nations*, Knoft, New York, 1991, and see R. Reich, 'Third way the hard way', *Age*, 12 October 1998.

2 Anthony Giddens, *The Third Way*, Polity Press, Cambridge, 1998, p. 26.
3 Lindsay Tanner, *Open Australia*, Pluto Press, Sydney, 1999, p. 15.
4 ibid., p. 8.
5 *Victoria: A Global Centre for the Information Age*, Multimedia Victoria 21, 1997.
6 *australia.com: Australia's Future Online*, Australia's Coalition of Service Industries, Melbourne, 1997, p. 32.
7 Terry Cutler, Deputy Chairman, National Office for the Information Economy Advisory Board, AIIA Forum, 12 August 1998.
8 D. Northfield, E. Richardson and T. Barr, *Any to Any Forum Report*, CIRCIT Ltd, August 1996, p. ii.
9 Sandy Kyrish, 'Public Policy and The Cycle of Prediction', BTCE Conference Paper, Melbourne, October 1996.
10 B. Dervin and P. Dewdney, 'Neutral questioning: A new approach to the reference interview', *RQ*, vol. 25, no. 4, 1986, p. 506.
11 Supriya Singh, 'Communication and information technologies', paper, CIRCIT, Melbourne, 11 August 1995.
12 *Review of the Standard Telephone Service*, DOCA, Canberra, 1996, p. 3.
13 ibid., p. 62, quoting R. Buckeridge, *Rural Australia On Line: Electronic Information Systems for Building Enterprises and Community Beyond the Cities*, Rural Industries Research and Development Corporation, Canberra, 1996, p. 35.
14 *Review of the Standard Telephone Service*, op. cit., pp. 179–80.
15 Australian Communications Authority, *Digital Data Review*, August 1998, p. 3.
16 ibid.
17 See Richard Butler, 'Evolving delivery models', *Telecommunication Journal of Australia*, vol. 48, no. 3, 1998.
18 'Protecting universal service obligations in a competitive telecommunications market', paper prepared by Telstra for the CTN Conference, Sydney, November 1998.
19 Broadband Services Expert Group, *Networking Australia's Future*, AGPS, Canberra, 1994, p. 52.
20 Terry Cutler, 'Call the bluff, this is the right time for us to go digital', *Age*, 30 March 1998.
21 'What is digital television?', *Fact Sheet*, Department of Communications and the Arts, 23 March 1998.
22 See Tony Branigan, General Manager of FACTS, 'Digital television policy', Seminar paper, Media and Telecommunications Policy Group, RMIT, Sydney, 9 February 1998.
23 ibid., p. 15.
24 Finola Burke, 'The ABC of digital TV', *Australian Financial Review*, 6 March 1998, p. 32.
25 Finola Burke, 'Telstra sees damage from TV gift', *Australian Financial Review*, 3 March 1998, p. 4.
26 Debra Richards, 'A fair go for digital broadcasting', *Communications Law Bulletin*, vol. 17, no. 1, 1998, p. 6.
27 Steve Lewis, 'Digital TV: Policy row', *Australian Financial Review*, 7 May 1998, p. 1.
28 ibid.
29 ibid., p. 8.

30 Finola Burke, 'Murdoch seeks licence to broadcast News', *Australian Financial Review*, 20 November 1998, p. 1.

9 Re-thinking our communications strategy

1 Barry Jones, 'The information revolution in Australia: Its impact on politics, the economy and society', address at the Sofitel Hotel, Melbourne, 14 April 1999, p. 4.
2 See two essays, 'Technical progress and the social life world' and 'Technology and science as ideology', in Jürgen Habermas, *Towards a Rational Society: Student Protest, Science and Politics*, trans. J.J. Shapiro, Heinemann, London, 1971.
3 W.H.G. Armytage, 'The rise of the technocratic class', in *Meaning and Control*, eds D.O. Edge and J.N. Wolfe, Tavistock, London, 1973, p. 65.
4 'Assembling the new economy', *The Economist*, 13 September 1997, p. 97.
5 *Investors Business Daily*, 10 June 1999.
6 *E-Commerce: Our Future Today*, Andersen Consulting, April 1998.
7 ibid.
8 Kevin Kelly, speaking at the Virtual Opportunities Congress, Parliament House, Melbourne, 29 October 1998.
9 E. Richardson, S. Miller and S. Singh, *Effective Use of On-Line Services*, CIRCIT Ltd, Melbourne, 1997.
10 Simon Molloy, 'The importance of connections', *Australian Financial Review*, 22 July 1998.
11 The term was used in 'Any to Any', CIRCIT Policy Forum, 12–14 June 1996.
12 See Wayne Hope, 'Whose All Blacks?', Paper, Auckland Institute of Technology, 1998.
13 Catherine Robinson, 'Australia and Convergence—Highway or Byway?', at http://www.gu.edu.au/centre/cmp/Papers_98?Robinson.htm
14 Stan Beer, 'SMEs cry foul over outsourcing decision', *Australian Financial Review*, 27 March 1999, p. 21.

Select bibliography

Allen Consulting Group Pty Ltd and Allen and Buckeridge Pty Ltd 1997, *'Spectator or Serious Player? Competitiveness of Australia's Information Industries'*, Report to the Information Industries Task Force, Canberra.

Anderson, B. 1983, *Imagined Communities: Reflections on the Origins and Spread of Nationalism*, Verso, London.

Anderson, S.J. 1996, *Policy Coalitions Rebuilding Competitiveness: Construction of Japan's Information Infrastructure*, Centre for Global Communications, Tokyo.

Ang, I. 1991, *Desperately Seeking Audiences*, Routledge, London.

——1996, *Living Room Wars*, Routledge, London.

Armstrong, M. and Molner, H. (eds) 1996, *Control and Ownership of Australian Communications*, Media and Telecommunications Policy Group, RMIT, Melbourne.

Australian Broadcasting Tribunal 1982, *Cable and Subscription Television Services for Australia* (Jones Report), 5 volumes, AGPS, Canberra.

Australian Communications Authority (ACA) 1998, *Digital Data Review*, ACA, Melbourne.

Australian Science and Technology Council (ASTEC) 1995, *Surf's Up: Future Needs 2010*, AGPS, Canberra.

——1995, *Future Needs 2010*, AGPS, Canberra.

Australian Telecommunications Authority (ATA) 1996, *Telecommunications Universal Service: Measures of its Delivery in Australia*, ATA, Melbourne.

Australia's Coalition of Service Industries (ACSI) 1997, *Australia.com: Australia's Future On Line*, ASCI, Melbourne.

Balnaves, M. and Richardson, E. 1990, *Social Equity and Telecommunications: The Applications of the Principles of Social Equity to the Telecommunications Area*, Strategic Analysis, Telecom Australia, Melbourne.

Bangemann, M. et al. 1994, *Europe and the Global Information Society: Recommendations to the European Council*, European Commission, Brussels.

Barr, T. 1985, *The Electronic Estate: New Communications Media and Australia*, Penguin, Melbourne.

——(ed.) 1987, *Challenge and Change: Australia's Information Society*, Oxford University Press, Melbourne.

——1995, *The New Zealand Equipment in the Liberalisation of Telecommunications*, Research Report No. 10, CIRCIT, Melbourne.

——1997, *Reflections of Reality: The Media in Australia*, Rigby, Adelaide.

——1997, 'U Turn or You Turn? Australia's IT Industry Development Strategy', *Telecommunications Journal of Australia*, Melbourne, vol. 47, no. 3.

Barr, T., Burke, J., Doddrell, M. and Northfield, D. 1995, *Foundations for the Future: Australian Telecommunications Equipment Industry*, CIRCIT, Melbourne.

Bauer, B. 1993, 'Challenges for German telecommunications policy: conference report', *Telecommunications Policy*, vol. 17, no. 1.

Bell, D. 1976, *The Coming of Post Industrial Society*, Heinemann, London.

——1996, 'The social framework of the information society', in *The Microelectronics Revolution*, ed. T. Forester, Basil Blackwell, London.

Benoist, A. 1996, 'Confronting globalisation', *Telos*, no. 108, Summer.

Broadband Services Expert Group 1994, *Networking Australia's Future*, AGPS, Canberra.

Broadcasting Research Unit (BRU 1998), *The Public Service Idea in British Broadcasting: Main Principles*, BRU, London.

Brown, A. 1995, 'Should Telstra be privatised?', paper presented at the 24th Conference of Economists, Adelaide, 24–27 September.

——1996, 'Reform and regulation of Australian telecommunications', Working Paper No. 5, Griffith University, Brisbane.

Bureau of Industry Economics 1995, 'International performance indicators: Telecommunications', Research Report 65, AGPS, Canberra.

Bureau of Transport and Communications Economics (BTCE) 1989, 'The cost of Telecom's community services obligations', Report 64, AGPS, Canberra.

——1993, 'International telecommunications: An Australian perspective', Report No. 17, AGPS, Canberra.

——1994, 'Demand projections for telecommunications services and equipment to Asia by 2010', Occasional Paper No. 109, BTCE, Canberra.

——1994, 'Elasticities of demand for telephone services', Working Paper No. 12, BTCE, Canberra.

——1995, 'Communications futures final report', AGPS, Canberra.

——1995, 'Interconnection pricing principles: A review of the economics literature', Working Paper No. 17, BTCE, Canberra.

——1995, 'Telecommunications in Australia', Report No. 87, AGPS, Canberra.

Button, J. 1998, *As It Happened*, Text Publishing Company, Melbourne.

Cairncross, F. 1997, *The Death of Distance: How the Communications Revolution Will Change Our Lives*, Harvard Business School Press, Boston.

Canadian Radio-Television and Telecommunications Commission (CRTC) 1995, *Competition and Culture on Canada's Information Highway: Managing the Realities of Transition*, CRTC, Ottawa.

Carey, J. 1989, *Communication as Culture: Essays on Media and Culture*, Unwin Hyman, London.

Castells, M. 1996–97, *The Information Age: Economy, Society and Culture* (series): *The Rise of the Network Society* (vol. 1, 1996); *The Power of Identity* (vol. 2, 1997); *End of Millennium* (vol. 3, 1997); Blackwell, London.

Centre for International Research on Communications and Information Technology (CIRCIT) 1992, *Research on Domestic Telephone Use*, proceedings of a CIRCIT-Telecom Workshop, CIRCIT, Melbourne.

——1995, 'Shaping the superhighway: Vision to reality', paper presented at 'What if Telstra is privatised?', CIRCIT Conference, Workshop No. 15, 28–29 September, Melbourne.

Commission of European Communities (CEC) 1987, 'Green paper on the development of the common market for telecommunication services and equipment', COM (87) 290 Final, CEC, Brussels, 30 June.

―――1993, 'Growth, competitiveness and employment—The challenges and ways forward into the 21st century', COM (93) 700 Final, CEC, Brussels.
―――1994, *Europe's Way to the Information Society: An Action Plan*, CEC, Brussels.
―――1994, 'Green paper on the liberalisation of telecommunications infrastructure and cable television networks', Part I, CEC, Brussels.
Committee of Review of the Australian Broadcasting Commission, report by, 1981, 'The ABC in review' (known as the Dix enquiry), AGPS, Canberra.
Commonwealth of Australia 1994, *Creative Nation*, Commonwealth Cultural Policy, AGPS, Canberra, October.
Communications Law Centre (CLC) 1998, 'Annual media ownership update', *Communications Update*, CLC, Sydney.
Compaine, Benjamin M., Weinraub and Mitchel J. 1997, 'Universal access to on-line services—an examination of the issue', *Telecommunications Policy*, vol. 21, no. 1, London.
Cunningham, S. 1998, 'Guidelines for an introduction to networking: A review of the literature', *The Arachnet Electronic Journal on Virtual Culture*, vol. 2, issue 3.
Cunningham, S. and Jacka, L. 1996, *Australian Television and International Mediascapes*, Cambridge University Press, Melbourne.
Cunningham, S., Jacka, L. and Sinclair, J. 1996, *New Patterns in Global Television*, Oxford University Press, Melbourne.
Curran, J., Gurevitch, M. and Woollacott, J. (eds) 1988, *Mass Communication and Society*, Edward Arnold, London.
Cutler and Company Pty Ltd 1995, *The Online Economy*, Cutler and Company, Melbourne.
Davis, G. 1997, 'The Mansfield Review: cautious, conservative and supportive', *Communications Update*, issue 130, March.
Department of Communications and the Arts (DOCA) 1994, 'Beyond the duopoly: Australian telecommunications policy and regulation', Issues Paper, AGPS, Canberra.
―――1996, 'Review of the standard telephone services', DOCA Canberra.
Department of Industry, Technology and Commerce 1987, 'An information industries strategy', *Australian Technology Magazine*, Canberra.
Department of Science, Technology and Industry 1997, 'Going for growth: Business programs for investment, innovation and export' (known as the Mortimer Report), APGS, Canberra.
Dervin, B. 1989, 'Users as research inventions: How research categories perpetuate inequalities', *Journal of Communication*, vol. 39, no. 3.
Dervin, B. and Dewdney, P. 1986, 'Neutral questioning: A new approach to the reference interview', *RQ*, vol. 25.
Dizard, W. 1997, *Old Media New Media: Mass Communications in the Information Age*, Longman, London.
Domberger, S. 1992, 'The role of public enterprise in microeconomic reform', in *Microeconomic Reform in Australia*, ed. P. Forsyth, Allen & Unwin, Sydney.
Domberger, S. and Piggott, J. 1985, 'Privatisation policies and public enterprise: A survey', *Journal of the Royal Statistical Society*, vol. 148, part 2.
Domberger, S. and Rimmer, S. 1994, 'Competitive tendering and contracting in the public sector: A survey', *International Journal of Economics and Business*, vol. 1, no. 3.
Dunleavy, P. 1992, *Democracy, Bureaucracy and Public Choice*, Prentice Hall, Englewood Cliffs, New Jersey.
Dutton, W. 1992, *The Social Impact of Emerging Telephone Services*, Telecommunications Policy, London, July.

Economist, The, 'Survey of telecommunications', 10 March 1990, 5 October 1991, 15 November 1993 and 30 September 1995.

Ergas, H. 1993, 'Privatisation and market forces: Their role in infrastructure provision', in *Economic Rationalism: Dead End or Way Forward?*, eds S. King and P. Lloyd, Allen & Unwin, Sydney.

—— 1994, 'An alternative view of Australian telecommunications reform', in *Implementing Reforms in the Telecommunications Sector: Lessons from Experience*, eds B. Welenius and P.A. Stern, World Bank, Washington, DC.

Evans, C. 1979, *The Mighty Micro*, Coronet, London.

Federal Bureau of Consumer Affairs 1997, *Untangling the Web: Electronic Commerce and the Consumer*, AGPS, Canberra.

Fidler, R. 1997, *MediaMorphosis*, Pine Forge Press, London.

Fielding, G. and Hartley, P. 1987, 'The telephone: A neglected medium', in *Studies in Communication*, eds A. Cashdan and M. Jordin, Blackwell, London.

Forester, T. 1985, *The Information Technology Revolution*, Basil Blackwell, London.

Forsyth, P. 1992, 'A perspective on microeconomic reform', in *Microeconomic Reform in Australia*, ed. P. Forsyth, Allen & Unwin, Sydney.

Gilder, G. 1994, *Life After Television*, Norton, New York.

Gillard, P. 1994, 'What do we really want? Rethinking media and telephone user research', *Media Information Australia*, AFTRS, Sydney.

Given, J. 1995, 'Australian content in communications hardware and software: Convergence on innovation', Communications Review Forum, 19 October, Sydney.

Green, L. and Guinery, R. (eds) 1994, *Framing Technology: Society, Choice and Change*, Allen & Unwin, Sydney.

Hamel, G. and Prahalad, C. 1994, *Competing for the Future*, Harvard Business School Press, Harvard.

Heap, N., Thomas, R., Einon, G., Mason, R. and Mackay, H. 1995 (eds), *Information Technology and Society: A Reader*, Sage, London.

Hilmer, F. et al. 1993, 'National competition policy', Report of the Independent Committee of Inquiry, AGPS, Canberra.

Hindle, J. 1997, *The Internet as Paradigm*, The Institute for Information Studies, a joint program of Nortel and the Aspen Institute, Nashville, TN.

Holton, R. 1998, *Globalisation and the Nation-State*, Macmillan, London.

Hudson, H. 1997, *Global Connections*, Van Nostrand Reinhold, New York.

Hutchinson, M. 1994, 'Telecommunications reform in Australia', in *Implementing Reforms in the Telecommunications Sector: Lessons from Experience*, eds B. Wellenius and P.A. Stern, World Bank, Washington, DC.

Industry Canada 1994, *The Canadian Information Highway: Building Canada's Information and Communications Infrastructure*, Industry Canada, Ottawa.

Information Highway Advisory Council 1995, *Connection, Community, Content: The Challenge of the Information Highway*, Industry Canada, Ottawa.

Information Industries Task Force 1997, *The Global Information Economy* (known as the Goldsworthy report), AGPS, Canberra.

Information Industry Advisory Council (IPAC) 1997, *A National Policy Framework for Structural Adjustment Within the New Commonwealth of Information*, IPAC, Canberra.

Information Infrastructure Taskforce (IIT) 1993, *National Information Infrastructure: Agenda for Action*, Executive Officer of the President IIT, Washington, DC.

—— 1994, *NII Progress Report*, IIT, Washington, DC.

Inglis, K. 1983, *This Is the ABC*, Melbourne University Press, Melbourne.

International Telecommunication Union (ITU) 1994, *World Telecommunication Development Report: World Telecommunication Indicators*, ITU Geneva.
——1995, *World Telecommunication Development Report: Information Infrastructures*, ITU, Geneva.
——1998, *World Telecommunication Development Report: Universal Access*, ITU, Geneva.
Jones, B. 1982, *Sleepers, Wake! Technology and the Future of Work*, Oxford University Press, Melbourne.
——1985, 'Australia as a post-industrial society', Occasional Paper, Commission for the Future, Melbourne.
Jones, C. 1996, *Telecommunications Universal Service: An International Perspective*, Austel, Melbourne.
Jordan, T. (ed.) 1998, 'Digital television policy', seminar proceedings, Media and Telecommunications Policy Group, RMIT University, Melbourne, 9 February.
Joseph, R. 1992, 'The politics of telecommunications reform in Australia and New Zealand: Lessons from a comparative analysis', *Pacific Telecommunications Review*, vol. 14, no. 2, December.
——1995, 'Direct foreign investment in telecommunications', *Telecommunications Policy*, vol. 19, no. 5, July.
Kapor, M. 1993, 'Where is the digital highway really heading?', *Wired*, July/August.
Katz, J. 1997, 'Birth of a digital nation', *Wired*, April.
Koelsch, F. 1995, *The Information Revolution*, McGraw Hill, New York.
Kuhn, T. 1970, *The Structure of Scientific Revolutions*, 2nd edn, University of Chicago Press, Chicago.
Langdale, J. 1995, 'International competitiveness on the information superhighway: Implications of East Asian developments for Australia', paper presented to the Communications Research Forum, Sydney, 19–20 October.
——1995, 'Issues in East Asian telecommunications industry development: Implications for Australian telecommunications policy', CIRCIT Discussion Draft, Melbourne, January.
Langtry, B. (ed.) 1998, *All Connected: Universal Services in Telecommunications*, Melbourne University Press, Melbourne.
Large, P. 1980, *The Micro Revolution*, Fontana, London.
Loader, B. 1997, *The Governance of Cyberspace: Politics, Technology and Global Restructuring*, Routledge, London.
Lovelock, P. and Ure, J. (eds) 1995, *The Broadcast Media Market in Asia: Telecommunications in Asia*, Hong Kong University Press, Hong Kong.
Lull, J. 1995, *Media Communication and Culture*, Columbia University Press, New York.
Lyon, D. 1998, *The Information Society: Issues and Illusions*, Polity Press, London.
Lyon, D. and Zureik, E. (eds) 1996, *Computers, Surveillance, and Privacy*, University of Minnesota Press, Minneapolis.
Maddock, R. 1992, 'Microeconomic reform of telecommunications: The long march from duopoly to duopoly', in *Microeconomic Reform in Australia*, ed. P. Forsyth, Allen & Unwin, Sydney.
Mansfield, R. 1997, *The Challenge of a Better ABC*, AGPS, Canberra.
Marien, M. 1996, 'New communications technology', *Telecommunications Policy*, vol. 20, no. 5, London.
Martin, B. 1993, *In the Public Interest?*, Zed Books, London.
Mayer, H. 1964, *The Press in Australia*, Lansdowne Press, Melbourne.

McQuail, D. 1992, *Media Performance: Mass Communication and the Public Interest*, Sage, London.

Media and Telecommunications Policy Group, 1998, *The Communications Environment 1998*, RMIT University, Melbourne.

Melody, W. 1986, 'Telecommunications: Policy directions for the technology and information services', *Oxford Surveys in Information Technology*, vol. 3, London.

——1990, 'Telecommunications: Policy directions for Australia in the global information economy', CIRCIT Policy Research Paper No. 7, Melbourne.

——1990, 'The information in IT: Where lies the public interest?', *Intermedia*, IIC, London, June–July.

——1991, 'Telecommunications reform: Which sectors to privatise?', paper presented at International Telecommunication Union Telecom '91 Economic Symposium, Geneva, 13 October.

——1996, 'Towards a framework for designing information society policies', *Telecommunications Policy*, vol. 20, London.

Ministerial Council for the Information Economy 1998, *Towards an Australian Strategy for the Information Economy*, National Office for the Information Economy, Canberra.

Moyal, A. 1985, *Clear Across Australia*, Nelson Publishing, Melbourne.

Multimedia Victoria (MMU) 1997, *Victoria: A Global Centre for the Information Age*, MMU, Melbourne.

Naisbitt, J. 1996, *Megatrends Asia*, Simon & Schuster, New York.

National Computer Board (NCB) 1992, *A Vision of an Intelligent Island: IT 2000 Report*, Singapore.

——1997, *Singapore One: Beyond the Internet*, NCB, Singapore.

Negraponte, N. 1995, *Being Digital*, Hodder, Sydney.

Newstead, A. 1993, *Telecommunications Innovations: Success Criteria*, Teltec Pty Ltd, Melbourne.

Nieuwenhuizen, J. 1997, *Asleep at the Wheel: Australia on the Information Superhighway*, ABC Books, Sydney.

Noam, E. 1992, *Telecommunications in Europe*, Oxford University Press, New York.

Noam, E., Komatsuzaki, S. and Conn, D.A. (eds) 1994, *Telecommunications in the Pacific Basin: An Evolutionary Approach*, Oxford University Press, New York.

Nora, S. and Minc, A. 1980, *The Computerisation of Society*, MIT Press, Cambridge, MA.

Norris, J.G. 1981, 'Report of the inquiry into the ownership and control of newspapers in Victoria', Report to the Premier of Victoria, Melbourne.

Northfield, D. 1994, 'Australian telecommunications commercial and regulatory environment, 1994', CIRCIT Research Report No. 8, November.

——1994, 'The role of public and private forces in telecommunications industry development: Experiences in Europe and Canada', CIRCIT Research Report No. 9, December, Melbourne.

——1995, 'Weighing up the arguments in the Telstra privatisation debate', unpublished paper, CIRCIT, 25 November, Melbourne.

Northfield, D. and Barr, T. 1995, 'Models of deregulation: An international comparison of telecommunications industry development policies', CIRCIT Policy Research Paper No. 37, March.

Office for Official Publications of the European Communities 1997, 'Effects on employment of the liberalization of the telecommunications sector', commissioned by the European Commission, Directorate-General V (Employment, Industrial Relations and

Social Affairs) and Directorate-General XIII (Telecommunications, Information Market and Exploitation of Research), Luxembourg.

Osborne, G. 1994, 'All roads lead to home or home is where the mart is: Information superhighways, human resources and the Toynbee dilemma: A research report', Centre for Communication Policy Research, University of Canberra.

Petre, D. and Harrington, D. 1996, *The Clever Country? Australia's Digital Future*, Lansdowne-Macmillan, Melbourne.

Pool, I. de S. 1983, *Technologies of Freedom*, Harvard University Press, Cambridge, MA.

Porter, D. (ed.) 1996, *Internet Culture*, Routledge, London.

——1997, *Internet Culture*, Routledge, New York.

Poster, M. 1994, 'A second media age?', *Arena*, no. 3, Melbourne.

Postman, N. 1993, *Technopoly*, Vintage Books, New York.

Randall, N. 1997, *The Soul of the Internet: Net Gods, Netizens and the Wiring of the World*, International Thomson Computer Press, Boston.

Reinecke, I. 1982, *Micro Invaders*, Penguin, Melbourne.

——1985, *Connecting You: Bridging the Communications Gap*, Penguin, Melbourne.

Reinecke, I. and Schultz, J. 1983, *The Phone Book*, Penguin, Melbourne.

Richardson, E., Miller, S. and Singh, S. 1997, *Effective Use of On-Line Services*, CIRCIT, Melbourne.

Rochlin, G. 1997, *Trapped in the Net*, Princeton University Press, Princeton, NJ.

Roszak, T. 1986, *The Cult of Information*, Pantheon Books, New York.

Rushkoff, D. 1994, *Media Virus*, Random House, Sydney.

Sardar, Z. and Ravetz, J.R. (eds) 1996, *Cyberfutures: Culture and Politics on the Information Superhighway*, New York University Press, New York.

Schiller, H. 1976, *Communication and Culture Domination*, International Arts and Science Press, New York.

——1996, *Information Inequality: The Deepening Social Crisis in America*, Routledge, New York.

——1997, 'Transactional media and national development', in *Media Culture and Regulation*, ed. K. Thompson, Sage, London.

Self, P. 1993, *Government by the Market: The Politics of Public Choice*, Macmillan, London.

Senate Reference Committee 1996, *Telstra: To Sell or Not To Sell?*, AGPS, Canberra.

Shoesmith, B. 1993, 'Technology transfer or technology dialogue: Rethinking Western communication values', *Media Asia*, vol. 20, no. 3.

Sinclair, J. 1994, 'Communication Media and Cultural Development', paper presented at Enhancing Cultural Value: Narrowcasting, Community Media and Culture Development, CIRCIT Conference, March, Melbourne.

Singh, S. 1995, *Communication and Information Technologies: The User's Perspective*, CIRCIT, August, Melbourne.

Sless, D. 1995, *Our Communication Ecologies: Paper for the Informationless Society*, Communication Research Institute of Australia, Canberra.

Slouka, M. 1995, *War of Worlds*, Abacus, London.

Smith, A. 1980, *Goodbye Guttenberg*, BBC video, London.

——1989, 'The public interest', *Intermedia*, vol. 17, no. 2.

——1992, 'Is there a global culture?', *Intermedia*, vol. 20, no. 4–5.

Smith, A., McQuail, D. and Held, V. 1970, *The Public Interest and Individual Interest*, Basic Books, New York.

Spender, D. 1995, *Nattering on the Net*, Spinifex Press, North Melbourne.

St Clair, Muir, J.J. and Walker, A. 1996, 'Digital technologies in Australian homes', paper presented at the Communications Research Forum, Melbourne, 28 October.
Stoll, C. 1996, *Silicon Snake Oil*, Pan Books, London.
Stone, A.R. 1995, *The War of Desire and Technology*, MIT Press, Cambridge, Mass.
Stonier, T. 1983, *The Wealth of Information: A Profile of the Post Industrial Economy*, Thames Methuen, London.
Sussman, G. and Lent, J. 1991, *A Transnational Communication: Wiring the Third World*, Sage, Thousand Oaks, California.
Tapscott, D. 1995, *The Digital Economy*, McGraw Hill, New York.
——1998, *Growing Up Digital*, McGraw Hill, New York.
Taylor, M. and Saarien, K. 1994, *Media Philosophy*, Routledge, London.
Telecommunications Council 1994, 'Reforms towards the intellectually creative society of the 21st century', Programme for the Establishment of High Performance Info-Communications Infrastructure Conference, Tokyo, May.
Tomlinson, J. 1994, 'Mass communication and the idea of a global public sphere', *Journal of International Communication*, July.
Toohey, B. 1994, *Tumbling Dice*, Heinemann, Melbourne.
'Top 500 companies' 1999, *Huntley's Shareholder*, 15th edn, Ian Huntley Pty Ltd, Sydney.
Turkle, S. 1995, *Life on the Screen*, Simon and Schuster, New York.
US Office of the Vice President 1995, *The Global Information Infrastructure: Agenda for Cooperation*, US Government Publishing Office, Washington, DC.
'Vision 21 for info-communications: Policies and practicalities for the 21st century' 1997, *New Breeze*, vol. 9, no. 3, July, Tokyo.
Wark, M. 1994, *Virtual Geography*, Indiana University Press, Bloomington.
Webster, F. and Robins, K. 1981, 'Information technology: Futurism, corporations and the state', *The Socialist Register*, eds R. Miliband and J. Saville, Merlin Press, London.
Wellenius, B. and Stern P.A. (eds) 1994, *Implementing Reforms in the Telecommunications Sector: Lessons from Experience*, World Bank, Washington, DC.
Were, J.B. 1998, *Australian Domestic Media Sector Report*, J.B. Were & Sons, Melbourne.
White, P. 1990, *Community Service Obligations and the Future of Telecommunications*, Commission for the Future, Melbourne.
White, R. 1981, *Inventing Australia*, George Allen & Unwin, Sydney.
Williams, R. 1974, *Television: Technology and Cultural Form*, Fontana, London.
——1984, *Towards 2000*, Penguin, London.
Wilson, R. and Dissanayake, W. (eds) 1997, *Global Local*, Duke University Press, North Carolina.
Windchuttle, K. 1984, *The Media*, Penguin, Melbourne.
Xavier, P. 1996, 'Australia's post July 1997 telecommunications regulation in international context', paper presented to the BTCE Communications Research Forum, 28–29 October, Melbourne.
——1997, 'Universal service and public access in the technologically dynamic and converging information society', a report prepared for the OECD, Melbourne.

Index

ABC
    ABC Enterprises, 63
    ABC TV, 63
    alleged bias of, 68–9
    audience share and ratings, 63
    contribution to Australian media sector, 70
    corporatisation of, 67
    critics of, 71
    establishment of, 61
    financial position, 64–5, 212
    friends and supporters of, 40, 70–1
    inquiries into, 67–8
    international television venture, 69
    pay television venture, 70
    position on digital television, 205
    program outsourcing by, 68
    as public broadcaster, 23, 66–7
    strategy towards new technology, 65
ACCC (Australian Competition and Consumer Commission)
    influence on telecommunications sector, 115
    interconnection decisions by, 95–6
    policy towards pay television, 58, 60
    as telecommunications regulator, 91–3, 115, 226
access
    to communications services, 148–9, 202–3, 220–2
    to new media, 147–50, 197–9
    to on-line data services, 197–203
    as universal and affordable, 102–5, 114–15, 148–9, 198, 202–3

Administrative Appeals Tribunal, 12
ADS-7, 11
Advertiser, 4, 11, 43
AFL (Australian Football League), 55
Age, 2, 4, 43
alliances, *see* mergers and alliances
Alston, Richard, 100, 108, 113, 192, 199, 201, 206
amazon.com, 138–40
America Online, 141
American National Football League, 44
American Sky Broadcasting, 44
Ang, Ien, 164
anti-competitive conduct, 92, 112–13, 114–15
APEC, 173–4
Argentina, 38
Asian financial crisis, 32, 33
AT&T (American Telephone and Telegraph), 26–7, 76
Aussat, 13, 48, 80, 82–3
Austel, 81, 90, 91, 97–8
Austereo, 3
Australia
    as an IT client state, 180–8
    communications policy for, 191–208 *passim*, 215–34 *passim*
    decline of electronics manufacturing in, 184–6
    history of telecommunications in, 78–86, 220–1
    lack of industry policy in, 187–8
    need to foster communications industry in, 174–6

Australian, 2, 4, 42, 43
Australian Broadcasting Authority, 5, 226, 229–30
  Internet regulation by, 133, 134–5
Australian Broadcasting Commission, 61–2, *see also* ABC
Australian Broadcasting Control Board, 5, 11
Australian Broadcasting Corporation, 67, *see also* ABC
Australian Broadcasting Tribunal, 5, 11–12, 14, 58
Australian Communications Authority, 91–3, 199–201
Australian Competition and Consumer Commission, *see* ACCC
Australian Consolidated Press, 3, 4, 47
Australian Financial Review, 4
Australian Home and Garden, 4
Australian Personal Computer, 4
Australian Postal Commission, 80
Australian Provincial Newspapers, 3
Australian Radio Network, 3
Australian Rugby League, 8
Australian Rules football, 55
Australian Subscription Television and Radio Association, 205
Australian Telecommunications Users' Group, 101, 115
Australian Women's Weekly, 2, 4
Australis Media Ltd, 58, 60

Bangemann report, 171
BBC, 61–2, 64, 161
Bell, Daniel, 167
Bertelsmann, 160
Black, Conrad, 52
Blair, Tony, xi, 189
Blount, Frank, 26, 111
Bond, Alan, 13, 50
Boston Herald, 44
British Telecom (BT), 26–7, 39
broadband services
  availability of, 202–3
  future types of, 197
broadcasting
  globalisation of, 160–1
  licenses for, 5, 10–16, 219–20
  ownership of, 23
  regulation of, 5, 11, 14, 58, 133–5, 226, 229–30
  social impact of, 5
  see also ABC; radio broadcasting; SBS; television networks
Broadcasting and Television Act, 5
  'Murdoch amendments' to, 11–12
broadcasting licences, 5, 10–16, 219–20
Brown, Allan, 100, 108
BT (British Telecom), 26–7, 39, 115
BTQ-7, 11
Buckeridge, Roger, 199
Buckeridge and Allen report, 181–2, 186
Bulletin, 2, 4
Bulletin Board Systems (BBS), 122
Business Review Weekly, 4
Button, John, 49

Cable and Wireless, 83
Cable and Wireless Optus, *see* Optus
cable television
  effect on commercial networks, 49, 155–66
  News Corporation's interests in, 45
Canada, 39
Canberra Times, 3, 54
CanWest, 56–7
Carlyon, Les, 1
Carroll, Vic, 7
casinos, 49
Castells, Manuel, 20, 32, 33
censorship, 119, 130–5
Channel 31, 73–4
Channel Nine, 9, *see also* ninemsn; Nine Network
Channel Seven, 3, 10, 11, 51, *see also* Seven Network
Channel Ten, 11, 13, 56–8
Chernin, Peter, 46
Chicago Sun Times, 44
Chile, 38
China
  regulation of Internet in, 132
  use of Internet by protest movements in, 122, 129, 163
cinema
  in Australia, 231
  in the Third World, 165
Clear Channel Communications, 3

Cleo, 4
Clinton, Bill, 129–30, 189–90
CNN, 161
Comcast, 27
commercial ideology, 7
commercial television
   concentrated ownership of, 11–16
   as oligopoly, 48
   response to digital television by, 205–8
   viewing of, 19
communications
   concept of, viii, ix
   Constitutional responsibility for, 5, 79
   convergence in, ix–x, 22–8, 168
   cyberculture model of, 125
   digitalisation of, 126
   economic significance of, 168
   network infrastructure for, 196–7
   new order in, 191–7, 215–17
   paradigm shift in, 144
   technological changes in, 22–31, 75, 167–8
   top-down broadcast model of, 125
   see also third communications order
communications industry
   composition of, 179
   convergence in, 22–5, 176
   economic significance of, 178–80, 215–16
   employment opportunities in, 218–19
   government promotion of, 169–76, 216–22
   size and growth of, 178–80
   top companies in, 177
communications orders, 191–2, *see also* third communications order
communications policy, 169–76
   Australia, 174–6
   blueprints for, 176–80, 215–34 *passim*
   decision-making in, 211–12, 218
   equity considerations in, 148–9
   Europe, 171
   Japan, 172
   lack of shared discourse on, 213–14
   South-East Asia, 172–3
   United States, 170–1
communications services
   disparity in access to, 147–8

   language employed in, 159–60
   network infrastructure for, 196–7
   user interactivity with, 158–60
   users' needs for, 195–6
community, notion of, 124–6
community radio and television, 72–4
community service obligations, 102–3
Community Telephone Plan, 79, 102
competition policy, 91, 218
   effect on pay television, 59–60
   effect on telecommunications, 58–61, 86, 90–3, 191, 212
   inadequate as IT&T framework, 210
   inconsistency of digital television policy with, 207–8
   interconnection requirements, 93–6
   outcomes, 101, 114, 233
competitiveness, 181–2
competitiveness, international, 181–2
computer corporations, 22–3
computing, and convergence, 22–5
Concert group, 26–7
consumers, *see* users
Consumers Telecommunications Network, 104
convergence in communications
   concept of, ix–x, 22–5, 176
   corporate effects of, 26–7, 160
   cross-advertising as effect of, 26
   implications for industry regulation, 225
   implications for 'old' media, 156
   multiplier effects of, 209–10
   new information services under, 225
   paradox of, 28
   response of News Corporation to, 46
Cosmopolitan, 4
Courier-Mail, 4, 43
cricket, 48
cross-media ownership rules, 14–15, 47, 52, 227
cross-subsidisation
   inequities in, 97
   of telephone network, 79, 96–7, 102–3, 149, 220–2
Crown Casino, 4, 49
cultural imperialism, 163–5
Cumberland Newspapers, 42

Cutler, Terry, 193–4
cyberspace, Internet as, 123–8

Daily Guardian, 2
Daily Mirror, 42, 43
Daily News, 43
Daily Telegraph, 4, 42, 43
data services, 197–203
Davidson, Kenneth, 8, 112
Davidson inquiry, 97
de Sola Pool, Ithiel, 166
decision-making
   in information policy, 211, 218
delivery capability, 28–9
deregulation, 38–9, 90–105 *passim,*
   109–10, 212, 230
Deutsche Telkom, 27, 39
digital television, ix, 203–8
digitalisation, 28, 77–8
Disney, 164
Dix inquiry, 67
Dolly, 4
Domberger, Simon, 109, 110
Dorfman, Ariel, 163
Duffy, Michael, 14, 49
duopoly, in pay television, 83–5, 92, 226

e-Commerce
   in Australia, 140–2
   defined, 135
   as economic challenge to
      Australia, 139–40, 142
   forms of, 136
   growth of, 136–7, 222
   as new commercial paradigm, 138–9
   potential of, 137, 223–4
   retailing as type of, 136–8
   ways to increase use of, 223
   see also amazon.com; Internet
      economy
economic rationalism, 38–9, 77, 109, 190
economy, *see* Internet economy
electronic books, 25
electronic commerce, *see* e-Commerce
electronic superhighway, 28
employment opportunities, 218–19
entertainment industry, global, 160
equitable access, 147–51, 202–3,
      220–2, *see also* access; community

service obligations; universal
service obligations
Ergas, Henry, 75, 199
ESPN, 161
Evans, Christopher, 146
Evans, Gareth, 11, 80–1

Fairfax, James, 19
Fairfax, Sir Warwick, 19, 51
Fairfax, Warwick, 51, 53
Fairfax family, 2, 51
Fairfax Ltd
   attitude to digital television, 205
   failed buy-out attempt, 51, 52
   financial performance of, 52–3
   Internet interests of, 53
   media controlled by, 2, 4, 11, 51–2
   ownership of, 51–2
   Packer group's interest in, 15, 47, 52
Fels, Allan, 60, 96
Fernback, J., 126
Festival records, 4
film industry
   Australia, 231
   Third World, 165
Financial Review, 43
Fox Entertainment group, 44–5
Fox Television, 26, 44
Foxtel, 3, 4, 22, 27
   ownership of, 50, 60
   pay television, losses of, 153
France Telecom, 39
Fraser government, 48–9, 80, 97
   and broadcasting licenses, 11–13
   community radio and broadcasting
      under, 73
Friends of the ABC, 70

gambling and gaming, 49
Germany, 39
Giddens, Anthony, 128, 189–90
Gilder, George, 154
Given, Jock, 9
'global village', *see* McLuhan, Marshall
globalisation, ix, 31–4, 147, 214
   effect on cultural identity, 161–2
   effect on nation state, 34–9
   efforts to control effects of, 33–4, 163
   of media and communications, 160–5

Goldsworthy report, 181–2, 186
Gore, Al, 170
government
 role of, 37–9, 76, 78, 215–34 *passim*
Gramsci, Antonio, 17

Habermas, Jürgen, 212–13
Hamilton, Clive, 209
Hawke government, 13–16, 49, 80
HDTV (high definition television), 204–5, 207
hegemony, 17
Herald (Melbourne), 2, 3
 circulation, 43
Herald Sun (Melbourne), 4
Herald and Weekly Times group, 2, 3, 10
 break-up of, 15
 takeover of, 42
Hewson, John, 209
Hilmer, Fred, 53, 227
Hilmer Committee, 91
Hollywood, 165
Home Box Office, 161
Hong Kong, 38, see also Star TV
Howard government, 187, 192
 attempt to censor Internet sites, 134–5
 policy on digital television, 205–8
 policy on media ownership, 15
 policy on R&D, 220
 Telstra privatisation by, 107–8, 110–11
Hoyts Cinemas, 4
HSV-7 television, 3

IBM, 26
ICT (information and communications technology), x
identity
 effect of Internet on notions of, 125
 effect of media globalisation on, 162
 impact of new notions of work on, 146
ideological diversity, 6–7
imperialism, see cultural imperialism
industry policy
 to foster local electronics manufacturing, 88–9

 lack of, 187–8, 212
information
 abundance of, 156–7
 concept of, 159
 digitalisation of, 28, 77–8
information-based economy, viii, 167
 Australian visions for, 192–4
 global aspect of, 32–3
 size of, 168–9
information networks, 29–31
information policy, 211–12
information revolution, 167
 effect on productivity, 157, 169
 emerging new services, 225
 involvement of users in, 154, 158
 key technological innovations in, 22–31
 over-optimistic predictions about, 152–3
 social impact of, 146, 233–4
 supply-led character of, 152
information society, ix, 20–1, 167
 inadequate public debate on, 210
 social impact of, 233–4
 visions for, 192–4
information technology, ix, 22–5
information technology industry
 declining manufacturing base of, 181–5
 inadequate local R&D in, 183, 212
 trade deficit in, 182
 weak venture capital market for, 182–3
 worsening international competitiveness of, 181–2
infrastructure
 for new communications order, 197–203
 for pay television, 59–60, 153
 public provision of, 76, 220–1
Innis, Harold, 128
interconnection, 93–6
Internet, ix, 22
 access to, 122, 150
 as alternative mode of expression, 122
 anarchic character of, 130, 143–4
 browsers for, 123, 126
 commercial opportunities of, 123

as contrast to top-down broadcast model, 124
effect on space/place notions, 127–9
effect on telecommunications manufacturers, 186
effect on television viewing, 19
as employment generator, 216
government services on, 232
growth and size of, viii–ix, 119–22, 185
as interactive cyberspace, 123–8
nature of, 118–19
new convergent services on, 225
as open communications medium, 118–19, 124–8
origins of, 120–2
as paradigm shift, 126–8, 143–4
political use of, 129–30
potential of, 224
psychological effects of, 127
regulation and censorship of, 130–5
  Australia, 130–5
  China, 132
  Malaysia, 133
  Singapore, 132
  United States, 119
rural access to, 221
service providers, 11, 122, 132, 140–1
strategy of Microsoft towards, 123
strategy of News Corporation towards, 46
users of, 119–20, 131
uses for, 118, 122, 126–7, 129–30, 131
see also e-Commerce; Internet economy
Internet economy, 222–4
  emergence of, 141–2
  size of, 136–7, 222
  see also e-Commerce
Internet hosts, viii
Internet service providers (ISPs), 11, 122, 132, 140–1
Internet stocks, 123, 141–2
ISPs, *see* Internet service providers
Italy, 39
IT&T (information technology and telecommunications), x, 210

IT&T networks, 28–31
IT&T policy, 211–12
IT&TI (information technology, telecommunications and Internet technologies), x

Japan, 38–9, 172
John Fairfax Holdings Ltd, *see* Fairfax Ltd
Johns, Brian, 66
Jones, Barry, 178, 211–12
journalistic conformity, 8–11

Kapor, Mitchell, 117, 124
Katz, Jon, 124
Keating, Paul, 1, 8, 14
Keating government, 49, 175–6
Kennett, Jeff, 193
Kuhn, Thomas, 117
Kyrish, Sandy, 152

Large, Peter, 146
Latham, Mark, 191
Leonard, Peter, 74
licenses to broadcast, 3, 5, 10–16
Lowy, Frank, 13
Lyons, Joseph, 10, 61
Lyons government, 10–11

MacCallum, Mungo, 9
Macdonald, Les, 33
Mahathir Mohamad, 33
Malaysia, 38, 133
manipulation, 18
Mansfield, Bob, 187
Mansfield inquiry, 67–8
Martin, Hans-Peter, 146
Marx, Karl, 16–17
mass communication, 20
mass media
  as contradictory term, 20
  as discredited term, 18
Mattleart, Armand, 164
McGuinness, Paddy, 8
MCI WorldCom, 26–7, 141
McKinsey and Company, 193
McLuhan, Marshall, 44, 128, 155, 162
media, ix
  abuse of power by, 6
  consumption of, 18

convergence with telecommunications and computing, 22–5
cultural imperialism of, 163–5
demassification of, 18, 227
globalisation of, 160–5
increased delivery capability of, 29
institutions of, 40–74 *passim*
lack of diversity of expression, 6–7
manipulation by, 18
participation by consumers in, 152
see also new media
media corporations, 40–61 *passim*
as conglomerates, 2, 6
mergers, 22, 26–8
takeovers of, 13, 15
see also media ownership and control; specific corporations
media institutions
commercial, 40–61 *passim*
public, 61–74 *passim*
media moguls, 1, 18, 49
Media One, 27
media ownership and control
concentration of, 3–4, 15, 17, 160
arguments against, 6–10
need to curb, 227–8
official inquiries into, 6
restrictions on, 12, 13–15
conflict of interests in, 7–8
cross-ownership of, 3
by foreign interests, 12
media policy, *see* broadcasting; cross-media ownership rules; two station rule
Menzies government, 11
Mercury, 4, 43
mergers and alliances, 26–8, 161
Mexico, 39
Microsoft
Internet strategy of, 123
investment in Web television by, 50
joint venture with PBL, 22, 27, 50–1
Mirror Newspapers, 41
mobile phones, 29, 98–9
carrier licenses, 83
use of and access to, 149, 152, 155
Web enabled, 24

Mortimer report, 182–3
MP3 players, 25
Murdoch, Elisabeth, 19, 46
Murdoch, James, 46
Murdoch, Lachlan, 19, 46, 207
Murdoch, Rupert, 2, 19, 40, 163
commercial unorthodoxy of, 46
global strategy of, 41, 45–6
heirs and successors to, 46
as newspaper owner, 12, 41–3
personal wealth of, 45
satellite-based media ventures, 44
as television network owner, 11–12
see also Murdoch family; News Corporation
Murdoch, Sir Keith, 2, 10, 19
'Murdoch amendments', 11–12
Murdoch family, 2, 6, 18, 19, *see also* News Corporation
Mushroom Records, 4

Naisbitt, John, 146, 153
nation states, 35
effect of globalisation on, 35–6, 214
national culture
effect of media globalisation on, 162, 164–5
National Information Services Council, 175
National Office for the Information Economy, 192–3
National Rugby League, 4
NCR, 26
Negroponte, Nicholas, 124
Netherlands, 39
Netscape, 123
networked society
concept of, 191
equitable access to, 220–2
possible broadband services for, 197
strategies for, 191–208 *passim*, 220–4
new communications order, *see* third communications order
new media
convergent services of, 24–5, 30–1, 225
implications for old media, 154–5
infrastructure for, 153–4, 197–203

language employed in, 159–60
uncertain demand for, 154
unequal access to, 147–50
user paradox, 156–60
New York Post, 44
New Zealand, 38–9, 230
News, 2, 43
News Broadcasting Australia, 207
News Corporation
   attitude to digital television, 205
   commercial unorthodoxy of, 46
   coverage of Super League, 9
   creation of Fox Entertainment group, 44–5
   as dominant Australian newspaper owner, 42–3
   globalisation of, 11, 160
   holdings of, 41
   Internet strategy of, 46
   investments in Foxtel, 3, 22, 27, 50
   investments in One.Tel, 26
   revenue of, 41–2
   satellite-based media ventures, 44
   software development by, 46
   US television interests of, 44–5
   vertical integration of, 46
   see also Fox Entertainment group; Foxtel; Fox Television; Murdoch, Rupert; Murdoch family; Star TV
News of the World, 43
newspapers
   acquisition of radio licenses by, 3
   circulation of, 19, 43
   family ownership of, 2, 19
   ownership of, 2–4, 6, 7, 9, 16, 17–19, 42–3
   time spent reading, 19
   see also cross-media ownership rules; specific newspapers and newspaper companies
Nine Network, 3, 4, 9, 13, 47, 50
ninemsn, 22, 27, 50–1
Northern Territory News, 4
Northfield, Dianne, 109
NWS-9, 11

Office of Film and Literature Classification, 131–2

oligopoly
   in Australian media, 2–3
   in press ownership, 2–3
Olivetti, 27
O'Neill, Helen, 189
One.Tel, 26
on-line data services, 30–1, 197–203
Onsale, 141
Optus, 50, 80, 83
   financial performance, 83–4
   involvement in pay television, 59–60, 153
   in mobile phone market, 84, 98–9
   purchases Aussat, 83
   stock market debut, 84
Optus Communications, 54
Optus Mobile Digital, 83
Optus/Telstra
   as duopoly, 83–5, 92, 226
   interconnection between, 93–6
Optus Vision, 59–60, 83
OTC (Overseas Telecommunications Commission), 80, 83
overseas telecommunications, 80
ownership, of media, *see* cross-media ownership rules; media ownership and control; newspapers
OzEmail, 140–1

Packer, James, 19, 49
Packer, Kerry, 2, 3, 8, 19, 46–51, 140
   advocates domestic satellite system, 80
   casino interests of, 49
   commercial strategy of, 47
   heirs and successors to, 49
   interest in acquiring Fairfax Ltd, 15, 47, 52
   Midas touch of, 49–50
   opposition to cross-ownership rules, 47
   personal fortune of, 48
   political influence of, 48
   sale of Nine Network to Bond, 50
   television holdings of, 47–8
   see also Packer family; Packer group; PBL
Packer, Sir Frank, 2, 19, 42

Packer family, 2, 6, 18, 19, 48
Packer group, 15, 47–8, *see also* PBL
paedophiles, 131
paradigm shift
  concept of, 117
  e-Commerce as, 138–9
  Internet as, 117, 126–8
pay television, ix, 3, 22, 58–61, 70, 149, *see also* Australis Media Ltd; Foxtel; Optus
  operating losses of, 59–60, 153
PBL (Publishing and Broadcasting Ltd), 6, 46–51
  casino stakeholdings of, 49
  financial growth of, 47
  holdings of, 4
  interests in Foxtel, 3, 50
  interests in Internet, 50–1
  interests in One.Tel, 26
  joint venture with Microsoft, 22, 27, 50–1
  media controlled by, 2, 4, 11, 51
  see also Packer, Kerry; Packer family; Packer group
PBL Enterprises, 47
PBL Online, 50–1
People, 4
personal computers, 147
Personal Investment, 4
policy
  on digital television, 205–8
  to foster domestic manufacturing, 88–9, 169–76, 187–8
  on information technology, 211–13
  Packer's influence on, 48
  on R&D, 220
  see also broadcasting; competition policy; regulation; telecommunications policy; universal service obligations
policy decision-making, 211, 218
political protest, 122, 129
post-industrial society, 146, 167
  Australia as, 178
Postman, Neil, 145, 156–7
power, of media, 6
powerlessness, 145
Powers, Brian, 53
press, *see* newspapers

price regulation, of Telstra, 97–102
privatisation, 105–6
  in Australia in 1990s, 106
  contrasted with deregulation, 109–10
  dilemmas of, 88
  of telecommunications carriers/operators, 38–9, 212
  of telecommunications companies, 105–6
  of telecommunications industry, 76–7, 88
  of Telstra, 88, 105–13, 115, 212
productivity, 157, 169
protest movements, see political protest
PTT, 39
public broadcasting
  by community radio and television, 72–4
  contrasted with public service broadcasting, 72–3
  critics of, 62
  rationale for, 61–2, 65–7
  role of ABC and SBS in, 66–7, 71–2
public interest, 61
public ownership
  rationale for, 102
  of telecommunications carriers, 76, 102
public sector, role of, 216–17
public service broadcasting, 61–72
Publishing and Broadcasting Ltd, *see* PBL

QTQ-9, 11
Queensland Press, 10
Quiggin, John, 111–12
Quintex, 13

radio broadcasting, 3, 11
Ravetz, Jerome, 146
R&D, 220, 223
regulation
  to achieve social goals, 102–5
  of broadcasting, 5, 11, 58, 133–5, 226, 229–30
  of Internet, 119, 130–5
  of media, 225–6
  need for single regulatory authority, 224–6

to promote children's programs on television, 229
to promote local content on television, 229
of telecommunications, 88, 90–3, 97–102, 225–6
of Telstra prices, 97–102
see also ACCC; Australian Broadcasting Authority; competition policy; cross-media ownership rules; deregulation; reregulation
Reich, Robert, 189–90
reregulation, 110
Rheingold, Howard, 125
Richardson, Graham, 8, 49
Rowley, Kelvin, 32
rugby league, 9
rugby union, 230
rural users, 79, 198–201, 221

San Antonio Express, 44
Sardar, Ziauddin, 146–7
satellite technology, 13, 44, 80, 161, 200, *see also* Aussat
SBS (Special Broadcasting Services), 23, 64, 71–2
Schumann, Harald, 146
Seagram, 160
Service Providers Action Network, 114
Seven Network, 13, 51, 53–6
Shawcross, William, 40
Skase, Christopher, 13, 51
Sky (satellite television), 44
Slouka, Mark, 127
Smith, Anthony, 20, 35
Smith's Weekly, 2
soap operas, 164
social goals, 102–5, 113–14, 198, 202, 221, *see also* universal service obligations
social impact
of information society, 146, 233–4
of Internet, 19, 127
software development
Australian opportunities in, 186, 224
by News Corporation, 46
Sony, 160

South-East Asia, 172–3
sports, broadcast of, 8, 9, 44, 48, 55
Staley, Tony, 48, 73, 81
Star TV (Hong Kong), 44, 161, 163
state, role of, 37–9, 76, 78, 215–34 *passim*
Stokes, Kerry, 53–6
Stoll, Clifford, 117
Stonier, Tom, 146
subscription television, *see* pay television
Sun (London), 2, 43
Sun (Melbourne), 43
Sun News Pictorial, 3
Sun (Sydney), 43
Sunday Age, 4
Sunday Herald Sun, 4
Sunday Telegraph, 2, 4, 42
Super League, 9
Sweden, 39
Sydney Casino, 49
Sydney Herald, 2
Sydney Morning Herald, 2, 4, 43, 51
Syme family, 2

takeovers, 13, 15
Tanner, Lindsay, 191
TARBS (Television and Radio Broadcasting Services Pty Ltd), 59, 61
technological change, xx, 22–31, 76, 184–6, *see also* convergence in communications; digitalisation; Internet
Telecom, 80, 83
cross-subsidisation of services by, 96–7, 103
Telecom Italia, 27, 39, 44
Telecom New Zealand, 39
telecommunications, ix, 22–5
conflicting policy objectives for, 86–90
Constitutional responsibility for, 5, 79
decline of local manufacturing in, 184–6
digitalisation of, 28, 77–8
future broadband services in, 197
government intervention in, 90–1

growth in, 152–3
historical development in Australia, 78–86
increasingly legalistic character of, 92–3, 115
overseas telecommunications, 80
policies to foster, 88–9
technological change in, 28–9, 79, 184–6
Telecommunications Act 1975, 80, 96
Telecommunications Act 1991, 85, 92, 94, 103
Telecommunications Act 1997, 91–2, 103–4
telecommunications carriers
deregulation of, 90–3
list of, 85
mergers and alliances between, 26–8
monopoly public ownership of, 76
open competition policy for, 85–7
ownership of, 23–4, 76
pricing policy for, 96–102
privatisation of, 76–7, 88, 105–13
regulation of, 81, 88, 90–3, 97–9
Telstra/Optus duopoly, 83–4, 85, 92, 226
see also Optus; Telecom; Telstra
telecommunications policy
pricing of services, 96–102
social goals of, 102–5, 113–14
see also universal service obligations
Telegraph (Queensland), 43
telephone network
cross-subsidisation of, 96–7, 102–3, 149, 220–2
government sponsorship of, 220–2
telephones
international use of, viii, ix
mobile phones, 24, 29, 98, 149, 152, 155
national telephony network, 198–203, 220–1
new telephony services, 151
ownership of
Australia, 102, 149
overseas, 147–8
trunk dialing, 79, 102
see also universal service obligations

television
children's programs on, 228–9
concentrated ownership of, 11–12, 48
digital, ix, 203–8
diversity of programs on, 157
globalisation in, 160–1
international disparity in access to, 147–8
local content on, 228–9
pay television, ix, 3, 22, 58–61, 70, 149, *see also* Australis Media Ltd; Foxtel; Optus
viewing of, 19
television licences, 3, 11–16
television networks
co-production alliances between, 161
licences for, 3, 11–16
as oligopoly, 48
ownership, 12, 14, 48, 227–8, *see also* cross-media ownership rules
Television and Radio Broadcasting Services Pty Ltd, *see* TARBS
Telmex, 39
Telstra
alleged inefficiency of, 109
anti-competitive conduct of, 93, 112–13, 114–15
creation of, 83
financial performance, 83–4
implications of digital technology for, 185–6
interconnection prices for competitors, 87
involvement in pay television, 3, 22, 27, 59–60, 153
market power of, 98, 115
in mobile phone market, 98
pricing of its services, 97–102
privatisation of, 88, 105–13, 212
staff retrenchment by, 87–8
universal service obligations of, 104
Telstra/Optus
as duopoly, 83–4, 85, 92, 226
interconnection arrangements between, 93–6
Ten Network, 13, 56–8
Test cricket, 48

Thatcherism, 77, 190
third communications order
   accessibility and affordability of, 197–8
   concept of, 191–2
   implications of digital television for, 203–8
   provision of infrastructure network for, 202
   role of public sector in, 216–17
   universal service obligations under, 197–203
   user focus of, 194–7
   visions for, 192–4
'third way'
   applied to communications policy, 191, 192
   concept of, xi, 189–91, 217
Thompson, B., 126
3DB radio, 3
Tiananmen Square protests, 122, 129
Ticketek, 50
Tiffen, Rod, 15
Time Warner, 160
Times (London), 2, 43–4
Toffler, Alvin, 153
Tomlinson, John, 31
Trade Practices Act, 81, 90–1, 130
transnational corporations, 36–7, 89
Turkle, Sherry, 125
20:80 society, 146
two station rule, 12, 14

United Kingdom, 38–9
United States, 169–76
'universal reach', 202–3
universal service obligations, 102–5, 114–15, 148–9
   and on-line services, 197–203, 199
user paradox, 156–60

users
   access to the new media, 147–50, 197–8
   interactivity with new services, 158–60
   lack of interactivity by, 154, 158–60
   needs of, 194–7
   in rural areas, 79, 198, 200–1, 221
   see also access; user paradox
USOs, *see* universal service obligations

venture capital, 182–3
Viacom, 160
Victorian government, 193, 232
videocassette recorders (VCRs)
   take-up rate for, 155
Village Roadshow, 3
virtual communities, 125
virtual reality, 127–8
visions, 189–208 *passim*
Vodaphone, 83, 98

Walker, Robin, 10
Walsh, Max, 9
Walt Disney, 160, 164
Ward, Michael, 208
web sites, *see* Internet
WebTV, 50
West Australian, 3, 43
Westerway, Peter, 8, 49
Wheels, 4
Wilkinson, Marian, 8
Woman's Day, 4
World Wide Web, ix, 118, 143

Xavier, Patrick, 99

Yahoo!, 26, 123, 141